I0755703

A New Sacred Geometry

The Art and Science of Frank Chester

by Seth T. Miller

To all curious souls.

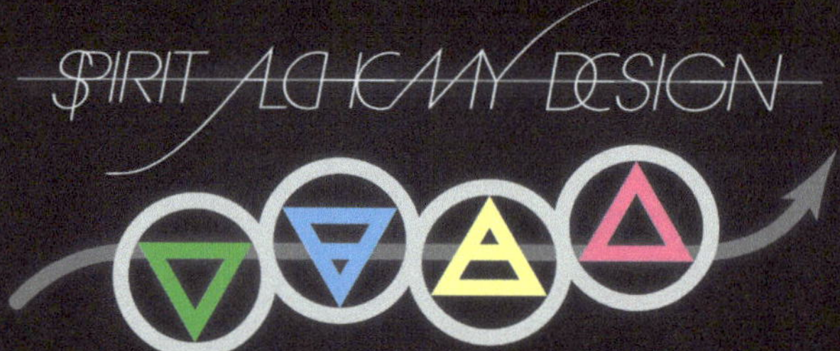

Photography: James Heath & Dana Rossini Rogers
Digital Artwork and book design: Seth T. Miller

Printed in the United States of America

ISBN: 978-0-9887492-0-7

Contents

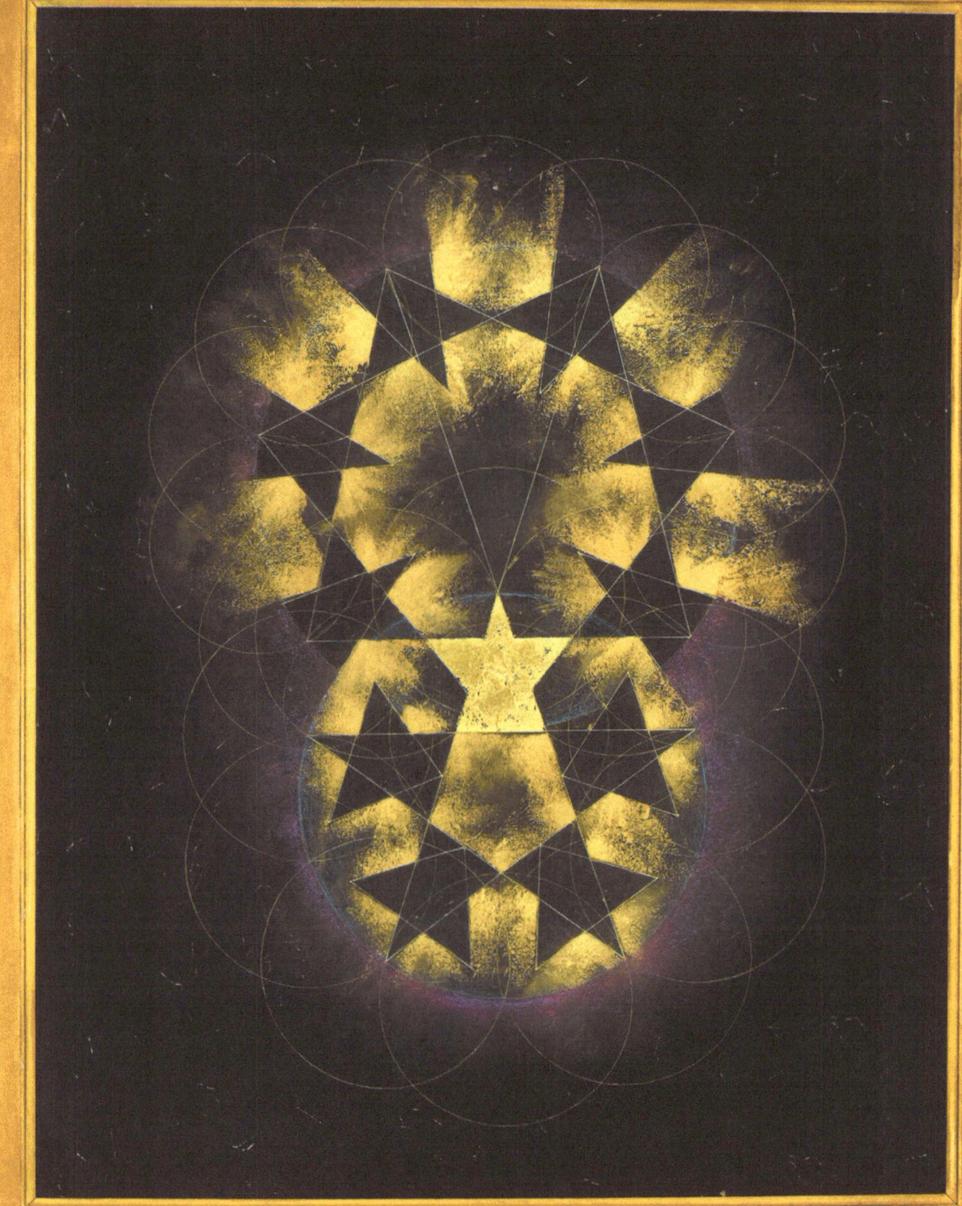

Foreward: Perspectives

Frank Chester is truly one of those rare geniuses whose work will surely transform the contours of science, art, and spirituality in the decades and centuries to come. His discovery of the Chestahedron, a new form that conceptually and geometrically unifies the five Platonic solids, newly extends over two thousand years of research by renowned scholars and scientists, and alone should secure his place in history. But to see the profound links between the Chestahedron and the form and dynamics of the human heart, the geography of the Earth, its potential applications in agriculture and environmental restoration, not to mention art and architecture as well, is to realize that one is standing before a fountain of deep discoveries and wisdom ready to bless humanity. I am excited to see Frank's work finally entering into the consciousness of the world.

Nicanor Perlas, Recipient 2003 Right Livelihood Award (also known as the Alternative Nobel Prize), author of the international best selling book, Shaping Globalization: Civil Society, Cultural Power, and Threefolding, *and Co-Founder, Movement of Imaginals for Sustainable Societies thru Initiatives Organizing and Networking (MISSION), www.imaginalmission.net.*

September 13th, 2012

On meeting Frank in his humble apartment in San Francisco, I was immediately struck by his workspace, which seemed to be a combination of laboratory, library, art studio, and geometric playground all in one—a new alchemical workshop, a crucible of sorts. The myriad objects, most tightly stored in glass-paned cupboards because the space is so small, had a powerful presence; but so did Frank's voice, steeped with an enthusiasm born from over a decade of continuous inquiry. As the afternoon progressed, one miracle after the other appeared from the treasure-chest of Frank's mind, heart, and hands, linking Universe, Earth and Humanity. As we conversed, new concepts were made visible like gently emerging snowflakes: structured, coherent, and ephemeral. I realized that I was in the presence of an inventor who had re-opened portals that had been habitually closed for hundreds of years. These portals had once been pathways to an intuitive perception of the harmony and unity of the world that the 17th century Enlightenment had chosen to bury. Frank has courageously stepped through these portals to explore the landscapes on the other side, and has rendered his experiences into real objects of geometry, art, sculpture, and architecture. He was as astonished as anyone at what has emerged: intimations that there may well be an exacting higher purpose to form and order, a coherence which can be made perceptible, and yet which also touches the sublime.

When Frank finally produced the bronze heptahedron (the seven-sided form he discovered, called the Chestahedron), I made a decision! I would support this spiritual scientist however I could. My commitment is specific to his research into the geometry that underpins so much of our natural, visible world. Since that afternoon, hung by the San Francisco sea fogs, I have embarked on a journey with Frank as one of his many colleagues in supporting the emergence of the geometry of spiritual forms. My inner sense is that these forms are accelerating and giving new potential to the future. Some of the forms no doubt will contribute to the foundation of a new spiritual technology.

It was when Frank rotated his model of the heptahedron and presented it in profile as the shape of a bell that my whole inner being moved. I could see this beautiful bronze bell hanging in the courtyard of a new college that I was undertaking to manifest with colleagues in Sheffield's city center. He called it the "Heart Bell." Yes! A Heart Bell in the center of Sheffield, a Heart Bell to be rung by students who are bereft and in a city whose foundations are built on metallurgy. A Heart Bell that would also be placed into existing schools formed by the Ruskin Mill project, crossing England from Northeast to Southwest. Each Ruskin Mill school would cast a bell and celebrate the transition of the seasons. Yes, this bell would radiate something completely new out of the old.

On the 29th September 2012, the new Heart Bell will be rung at Clervaux in Darlington, adjoining the river Tees, and a thread of sound will ring in the Michaelmas event. More than this, the smaller bells, a "Heart Family," will also in time be ringing out across England from Cambridge Rudolf Steiner School to Plas Dwbl in Pembrokeshire, Wales. I intend to avail this Heart Bell in countless social, educational and biodynamic projects.

I perceive Frank's contribution to be the starting point of a new future in which living geometry can contribute to and help shape an emergent culture. Within this geometry, it is possible to start to perceive how archetypes and meta-principles linking the universe and humanity can be utilized in a conscious application toward human and spiritual development.

Aonghus Gordon M.Ed., founder and director of the Ruskin Mill Trust (www.rmt.org), which works with challenged young people who undertake a three-year training program in Practical Skills Therapeutic Education. There are currently six centres throughout the UK. The work is inspired by Rudolf Steiner's spiritual psychology, craft work and the natural world. In 2005 Aonghus was awarded the UK Social Entrepreneur of the Year.

September 23rd, 2012

Setting One – The Tradition of Sacred Geometry

The small group of sacred geometers has a special language. They speak in hushed tones of squaring the circle and circling the square, a medieval problem about how to equate the perimeter of a square—symbolizing the manifest earth with its rectilinear lines made necessary by gravity and ease of construction—with the perimeter of a circle—symbolizing the infinitude of heaven, from where the creations emanate. They seek a perfect solution, a mathematical solution or, even more elegant, a geometrical solution. The side of a square can be measured, but the circle has that unknown factor—pi—called a transcendental number because it transcends physical reality, and whose value can never be known exactly. When you accomplish this construction, you participate in the act of creation.

Sacred geometers nod and sigh when contemplating the equilateral triangle.

Most people do not comprehend the satisfaction experienced by geometers when simply gazing upon this form. The geometers see more than others. They realize that, before three references, there were two points, giving only a line, one dimension. Add a third reference point and you open from the restrictions of the line into the plane with a sigh of freedom in expansion, the equal distance between each point giving a kind of perfection to the relationships. Implied is the opening to the next dimension (the third - space), and the stability of the tetrahedron:

Buckminster Fuller re-discovered the power of this form and from his work we have domes and other forms that soar into a third dimension made freer by this most elegant of forms multiplied.

Sacred geometers smile when they see what have come to be called the Platonic solids: the four-sided tetrahedron, the eight-sided octahedron, the twenty-sided icosahedron—the ones whose sides are made of equilateral triangles —the six-sided cube—whose sides are made of squares —and the twelve-sided dodecahedron—whose sides are made of five-sided pentagons. Though described by Plato, the British Museum has stones carved in these perfect forms originating from thousands of years before Plato.

Sacred geometers nod when they see five-fold geometries, including the US military's Pentagon, because they see intrinsic to it $\sqrt{5}$, in the form of the golden mean or phi = $(\sqrt{5} + 1)/2$, the proportion of growth in flowers and vegetables and nautilus shells and many other forms—the process of growth copied into the temples of Egypt and the cathedrals of Europe. The square-root of 5 comes from the root (as in tree root, that is, the foundation and essential support) of the square whose value is 5. This is 5 of what? 5 dollars? 5 yards? 5 chickens? To a geometer the unit doesn't matter, for the proportion is so much more important—proportion and therefore angle ... as "the angles are angels." This is one of the secrets that sacred geometers hold—that there is an entire bandwidth of spiritual refinement made entirely of geometrical relationships. As matter densifies, these relationships become hidden, though they remain at the root of all forms. As you refine away from density toward source, you have to go through that bandwidth: Then sacred geometers become your guides. Astrologers love these angles as angels of square, triangle, quintile (from the pentagon), septile (from the seven-sided figure, which you will soon learn to construct and admire).

Sacred geometers nod when they listen to music, as they follow the lead of Pythagoras to hear the purity of the octave (fretting a vibrating string at exactly one half its distance), the perfect fifth (from the foundation *do* to *sol*, exactly 2/3 of the string - 2/5 and 3/5), and many other intervals and overtones. They nod because they see the same intervals in nature and in architecture and in everything that pleases the eye.

One form that especially delights a geometer is the vesica piscis, created when you put the center of a second circle on the perimeter of a first circle.

The central form, shaped as a vertical football, is traditionally termed the vesica piscis (bladder or bag of the fish), also the mandorla (shaped as an almond). In many traditions this form without the outer circles can be found to frame Mother Mary inside or Buddha inside or any of the Hindu gods inside. Even though the inner shape is shown without the construction elements, the geometer can see where it came from, and know that it relates to larger forms of intersecting circles, called the Flower of Life.

Sacred geometers nod knowingly because they feel, as do alchemists of old and modern scientists and also philosophers, that they have a special key to creation - to the earth, to Gaia - from which the term geometry comes—geo/Gaia, -metry/measure. William Blake pictured God with a large compass measuring out the earth and everything in it - in proportions that were and are by definition divine because God made them. Thus the School of Chartres has become known to the informal guild of sacred geometers as a gathering place for those with these interests - the training ground in the twelfth-century (when it flourished, though it began earlier) of sacred geometers. There have been other schools and centers, though few have taken these forms more seriously than ornament.

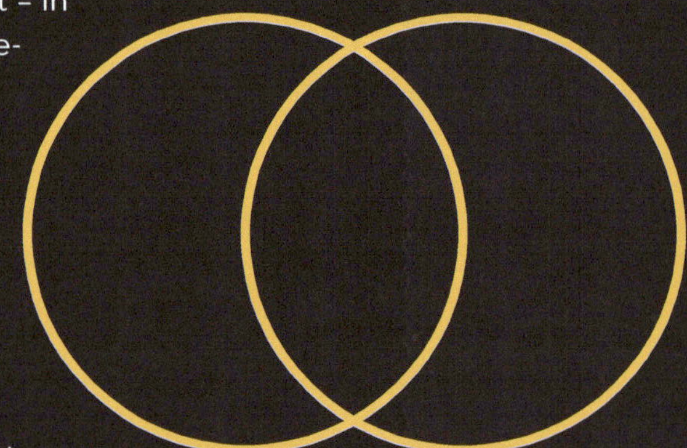

Those who designed and built the StarHouse in Boulder, Colorado, studied sacred geometry with Robert Lawlor (author of *Sacred Geometry*, first published in 1979, and selling several thousand copies each year since then) and with Keith Critchlow (*The Hidden Geometry of Flowers*). For nearly two years, we gathered as a group to study how to apply these principles to the StarHouse and the many features in the landscape around it. Such attentions resulted in an architecture that nourishes rather than depletes (and whose history and geometry can be found at www.TheStarHouse.org and at www.StarWisdom.org).

Setting Two - the Tradition of Anthroposophy

The Austrian philosopher Rudolf Steiner (1861-1925), responsible for what we see today in Waldorf schools, biodynamic agriculture, the art form of eurythmy, a new approach to medicine, and many other initiatives, began his teaching work just after the year 1900. Before that, he had edited the scientific works of Johann Wolfgang von Goethe, a great statesman, scientist, and playwright (*Faust* being the best known). By 1913 Steiner had a large following, and had just separated from theosophy to create anthroposophy, from anthropos (the possible human being) and Sophia (divine feminine wisdom). He decided to build a gathering place for the teaching of anthroposophy, through the performances of sacred drama (his own "mystery plays" as well as those by symbolists attracted to anthroposophy) and through his lectures (of which he gave sometimes four a day in different fields of endeavor). This building came to be known as the Goetheanum, and was designed as the merging of two dodecahedrons.

Steiner attracted many donations of money and materials. Large timbers of rare trees were sent from every continent to create the building. Many of the creative people of his day from many countries were lured there. Inspired young people who desperately needed to create rather than destroy—they could sometimes hear the artillery of World War I in the distance—came to this hillside in Switzerland to camp out and learn and work hard.

One of the tasks was to carve the tops of the seven great pillars that held the roof aloft in the larger dodecahedron. These were the capitals—the heads—of those columns, one for each of the planets. One woman wrote about the work:

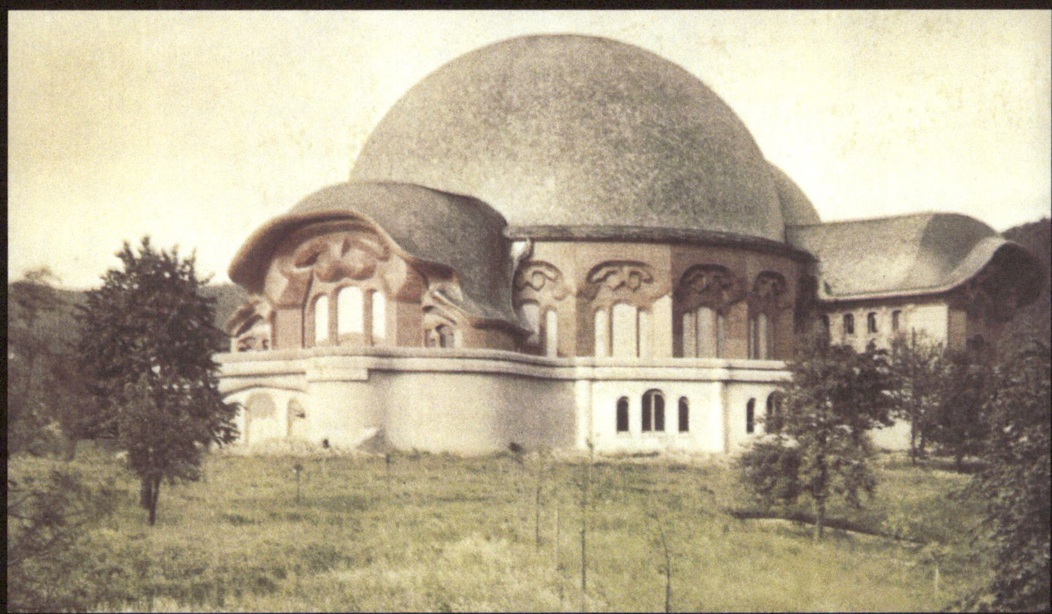

The first Goetheanum.

> We have before us a row of wooden blocks more than a yard each in height and width. ... We take chisels and mallets. A great effort is needed to cut even a splinter out of the block. We take counsel, we pull on the wood, we are already tired. Our arms are sore, yet one sees no result of the effort. Dr. Steiner ... takes a mallet and a chisel, climbs on a wooden crate and begins to work. He too has never done such work before, but after a few blows with the mallet he appears to be quite familiar with it and cuts one furrow after another. We see him hammer for ten minutes, one hour, two hours, without stopping. We stand at a distance, pale with exhaustion, and look at him in awed silence. We knew by experience how hard the work was.

These capitals were key to Steiner's gift to humanity - his understanding of the construction of the universe - his sacred geometry. One of the workers wrote a little play about the experience that included the following conversation about the capitals:

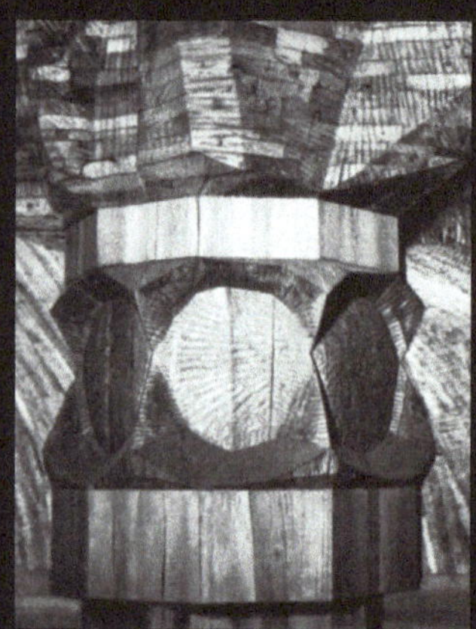

The first capital.

Overseer:

When will peace come?

Rudolf Steiner:

Only when the dead
can be included in our conferences.

Overseer:

But how?

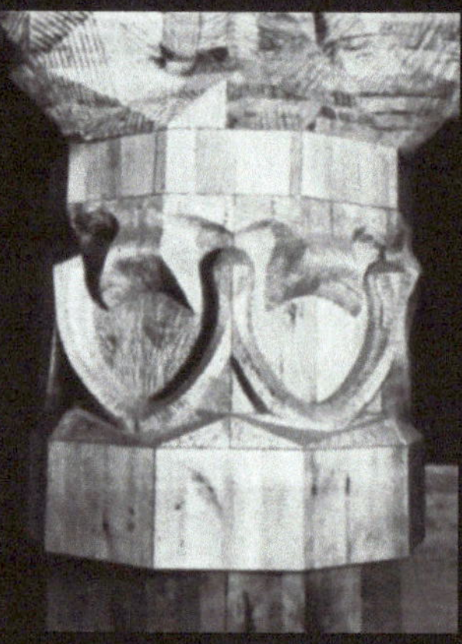

The sixth capital.

Rudolf Steiner:

Why, in the language of these forms.
Would that the living too might listen to them.
Then they would know themselves to be immortal,
while still encumbered with an earthly frame,
and cease to harbour hate in scornful hearts.

(You might return to these words after studying Frank Chester's work.)

One young person summarized these months in the following way: "He turned us from impractical, dreamy eccentrics who often ran away from life..."—one could almost read this as "hippies"—"into workmen who used their bodies in the service of the spirit." (All references for these quotes can be found in my book, *Star Wisdom & Rudolf Steiner: A Life Seen Through the Oracle of the Solar Cross*, from www.StarWisdom.org.)

That first Goetheanum burned to the ground on December 31, 1922. But, important to this narrative, its structures had been photographed.

Setting Three - Enter Frank Chester

Frank Chester taught art in high school and college for many years. This gave him great familiarity with materials and techniques, how to make ideas into objects, how to manifest in material form interpretations of creation, that is, art.

Upon retiring from his responsibilities to the public school system, Frank Chester found his way to Rudolf Steiner College and was promoted there by the wonderful teachers Dennis Klocek, Brian Gray, and Patricia Dickson. He studied Steiner's work, and was drawn to the Goetheanum. He found his way to the capital of the pillar for Saturn—no longer available as a three-dimensional reality, but through two-dimensional photographs. He thought he saw in it the revelation of a Platonic seven-sided figure. Recall that the Platonic solids include figures with sides of four, six, eight, twelve, and twenty sides; the notion of seven sides had long been sought for the power of the number of seven, but the seekers had given up. Frank created a series of models, some of ping-pong balls and wire, and some of much more sophisticated materials. He found such a figure.

Encouraged by his allies at Rudolf Steiner College, he put this sequence of discovery on display in the foyer of the main meeting hall for the "Anthroposophy in a New Millennium" conference at Rudolf Steiner College in June 2000. Hundreds of people attended. In the midst of droves of people coming and going, I passed by the exhibits by Frank Chester three times until I was by myself and finally was able to look more closely. I stopped dead in my tracks. Here were shown the roots (meaning the foundations, the essences) of what had been used to build the StarHouse. Here were the struggles of one man with the enigmatic capital of Saturn. And he had come out of this struggle with something absolutely unique and new and wonderful and yet recognizable. I looked him up and we have been friends ever since.

As an indication of how specialized the band of sacred geometers, Frank said that only two people contacted him on the basis of his display.

One more glimpse from this setting of Frank Chester. He does not have a large studio, but rather a dedicated corner of his apartment in which to work. Every time I visit, there is a different thing laid out. Sometimes there are pages of drawings, for he uses the classical methods of deriving the

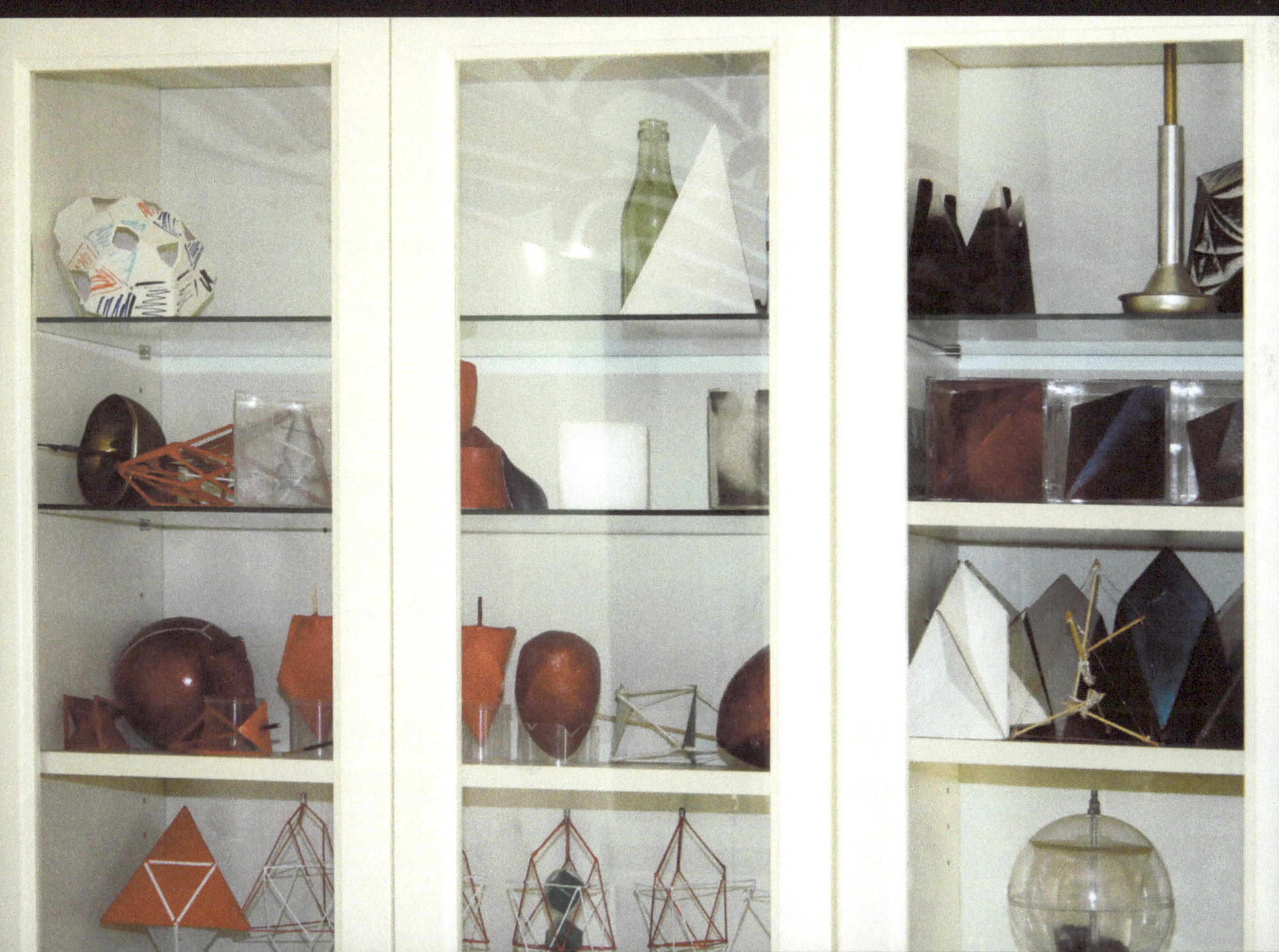

designs using the simplest of technologies—the compass. (The equations at the end of this book come into play also, but they are actually less important than the construction methods derived from traditional compass work.) Sometimes there are new constructions demonstrating how abstract shapes in movement create expansion/growth and decay/ vacuum. Once there was a large cylinder, six feet high and two feet in diameter, full of rotating water demonstrating the power of vortices to increase pressure and vacuums. Cosmologists speak conceptually about black holes and worm-holes through time and space, but Frank Chester can demonstrate them through geometrical elements in movement.

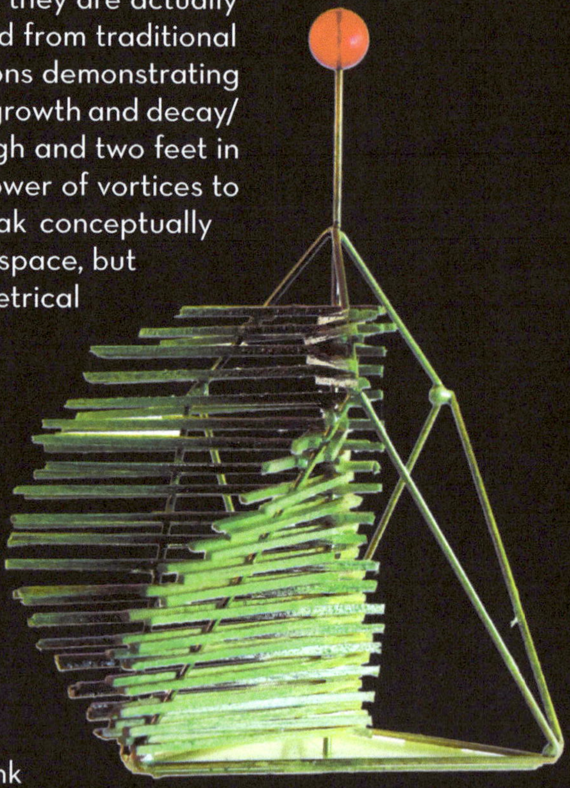

Summary

Frank Chester is surely a reborn master of the School of Chartres. He has taken the cue from Rudolf Steiner to create masterpieces of elegant geometry, and also to make them available through many fields of endeavor—from water purification to cathedral design to the dynamics of the human heart to the shape of sound (his bells) and ... one never knows what this master will create next. As the playwright had Steiner say, "Would that the living too might listen to [these forms]," so may it be with Frank Chester, sacred geometer.

David Tresemer, Ph.D., co-founder of the StarHouse in Boulder, Colorado (www.TheStarHouse.org), co-founder of the Star Wisdom approach to astrology (www.StarWisdom.org), author of The Venus Eclipse of the Sun 2012, *and* One-Two-ONE: A Guidebook for Conscious Partnerships, Weddings, and Rededication Ceremonies *(with Lila Sophia Tresemer), and other books, and co-founder of Mountain Seas intentional community (www.MountainSeas.com.au)*

September 9, 2012

My first meeting with Frank Chester was in the year 2000 in Fair Oaks, CA where he shared his research methods and discoveries with a small group of artists. In the previous months he had discovered the seven-sided solid composed of surfaces having two shapes and all having the same equal area, now called the Chestahedron. It was fascinating to hear the descriptions of his process and how each step led to the next. I was deeply moved by his enthusiasm and his modesty. The forms Frank produced were not merely simple creations, but gifts that he received through deep conversation with the principles of geometry. He asked questions and answers came. From my own studies of geometry and experience of the unique relevance of the number seven, the extreme importance of this occasion filled me with awe and reverence. Without knowing what was to come I knew the implications were enormous. Yet no one—not even Frank—could foresee the volume and potential applications of the forms to follow. When Frank revealed the Venus form, which appears like two embracing hearts I was astounded. This amazing curvilinear form had evolved through pure geometry as an inversion traced through time of the seven-sided form. Without any scientific evidence and based on its pure beauty and my intuition I believed the form to carry therapeutic value. I arranged for two commissions to be made in bronze. One was installed in a space for meditation and color/sound therapy in Kansas City, Kansas and the other was placed at the Noah Center in Great Barrington, MA, a center for education about alternative healing founded by Linda Norris.

I have attended a few other presentations of Franks work over the years and finally attended a workshop at Camphill Ghent, NY where Frank led us through the process of making the seven-sided form and using this to then arrive at a bell form, the first ever bell form derived from sacred geometric principles. These bells now exist in bronze in England and the United states. Frank's presentation of his most recent discoveries were quite complex and illustrate the geometry of the heart including the eight layers of the heart's muscles. There seems to be no end to what is being revealed through this research. Each geometric revelation has led to numerous others. Frank is on an adventure, pioneering in a new land.

Robert Logsdon, artist and master lazure painter, founder of ColorSpace: color consulting, design, mural and lazure color applications. Graduate of Cincinnati Art Academy and Waldorf teacher training, Emerson College, England.

October 4, 2012

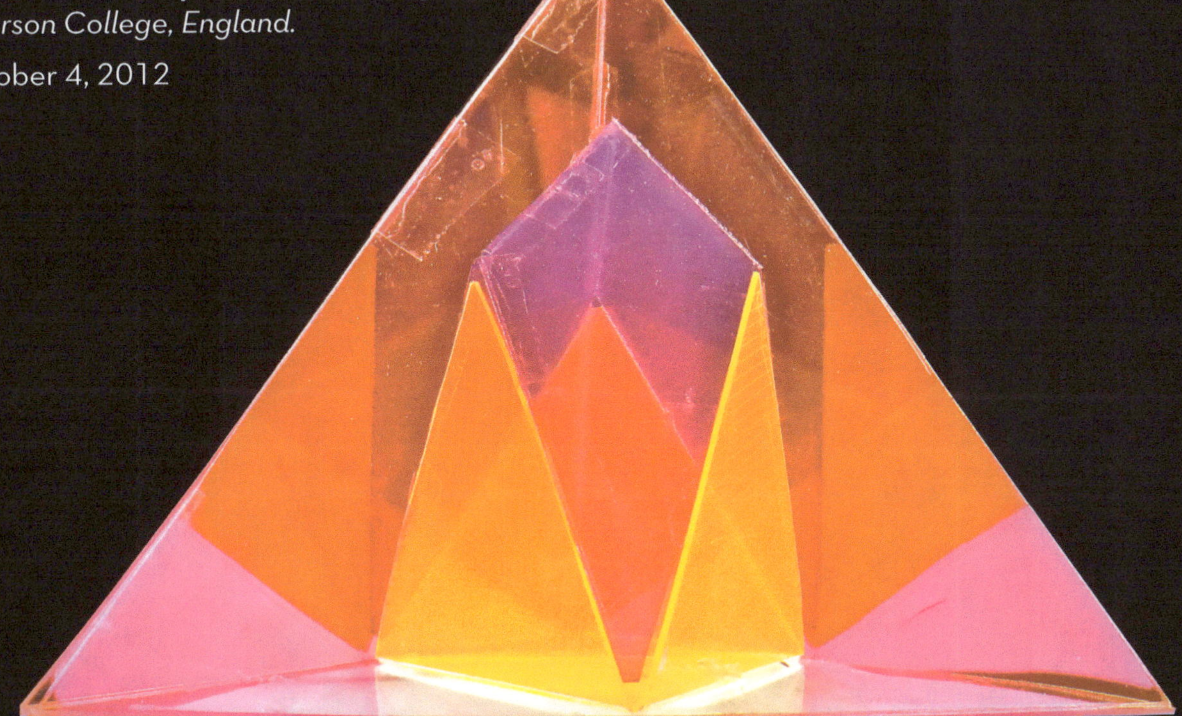

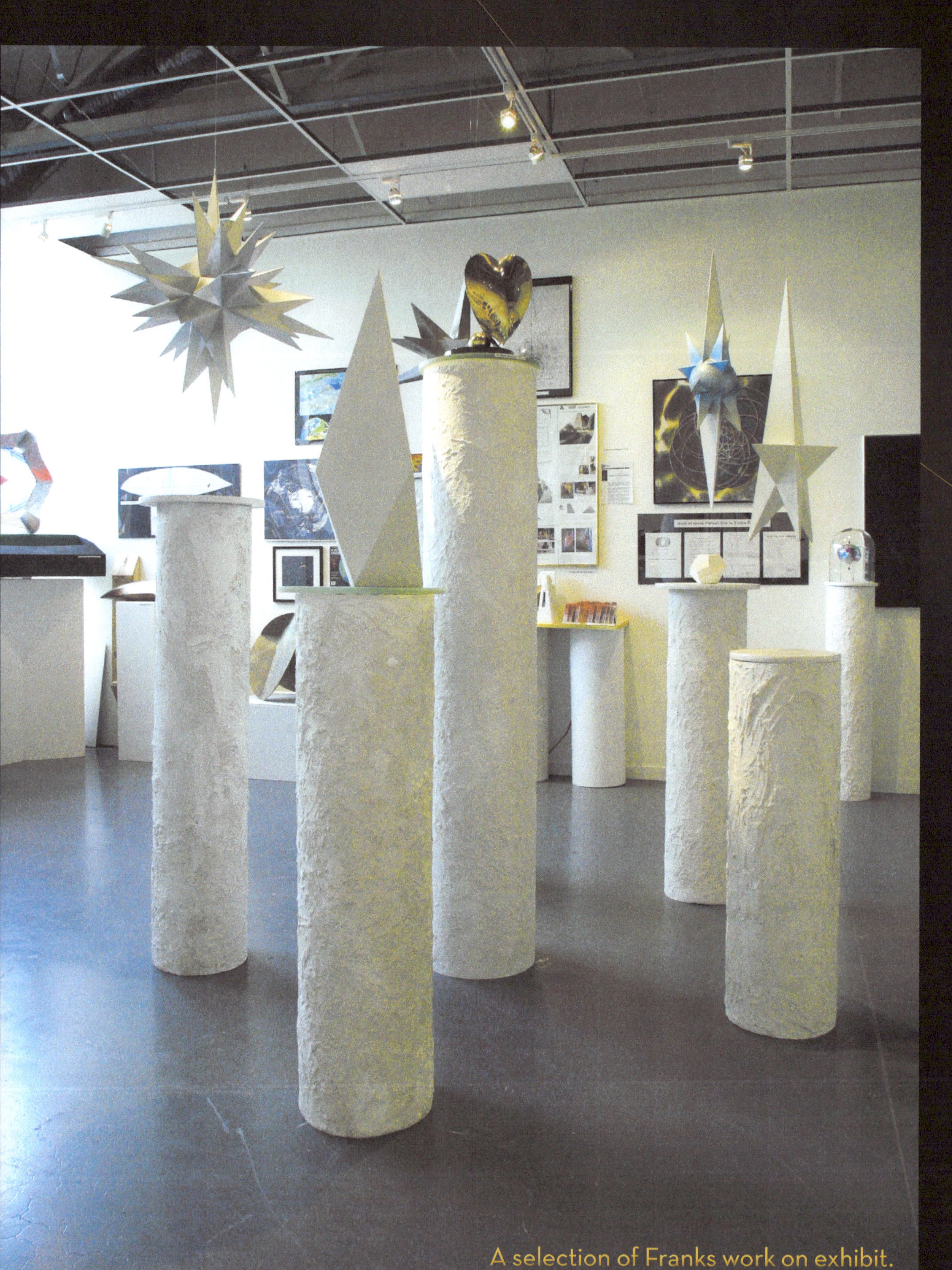

A selection of Franks work on exhibit.

The CHESTAHEDRON

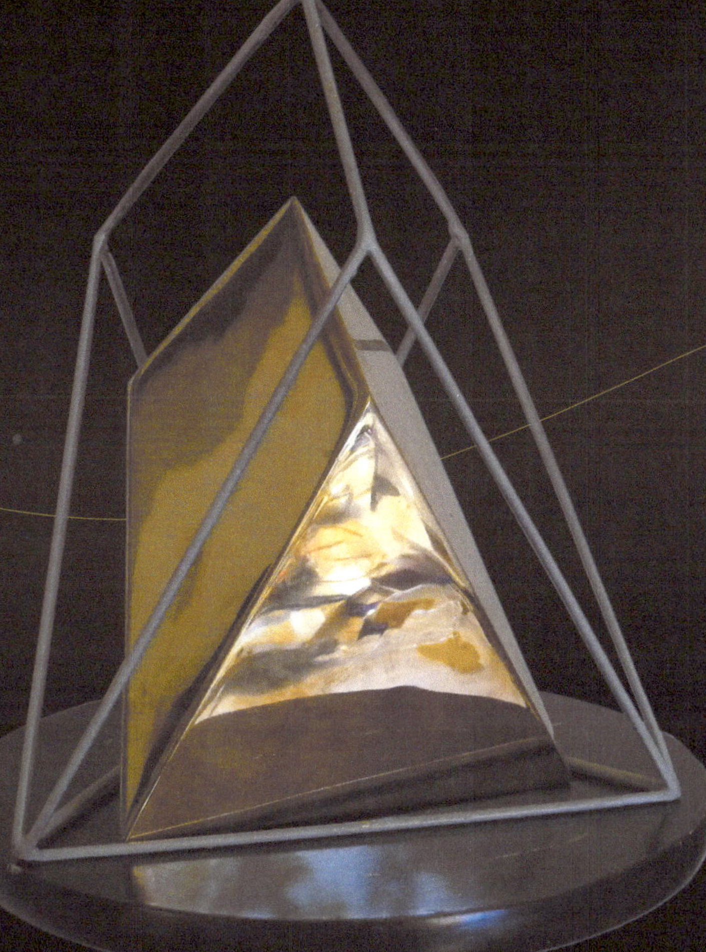

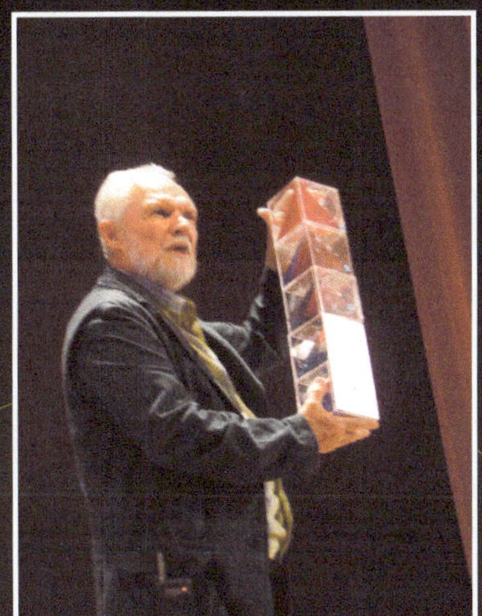

Frank presenting.

The Chestahedron is a new geometric form discovered by Frank Chester. It has seven faces of equal area, is comprised of four equilateral triangles and three quadrilaterals, and has a surprising amount of unique geometric properties. Frank's research has revealed a number of promising potential areas of application of this form, from architecture, vortexial mixing, beehive construction, bell-making, and the interior structure of the Earth, to the inner geometry of the human heart, not to mention its purely aesthetic qualities.

The Geometry of the Chestahedron

The Chestahedron is a three-dimensional solid newly discovered by Frank Chester in January 2000. It is a 3-fold rotational prismatic symmetrical heptahedron. The Chestahedron is the first known seven-sided solid with this configuration with faces of equal area.

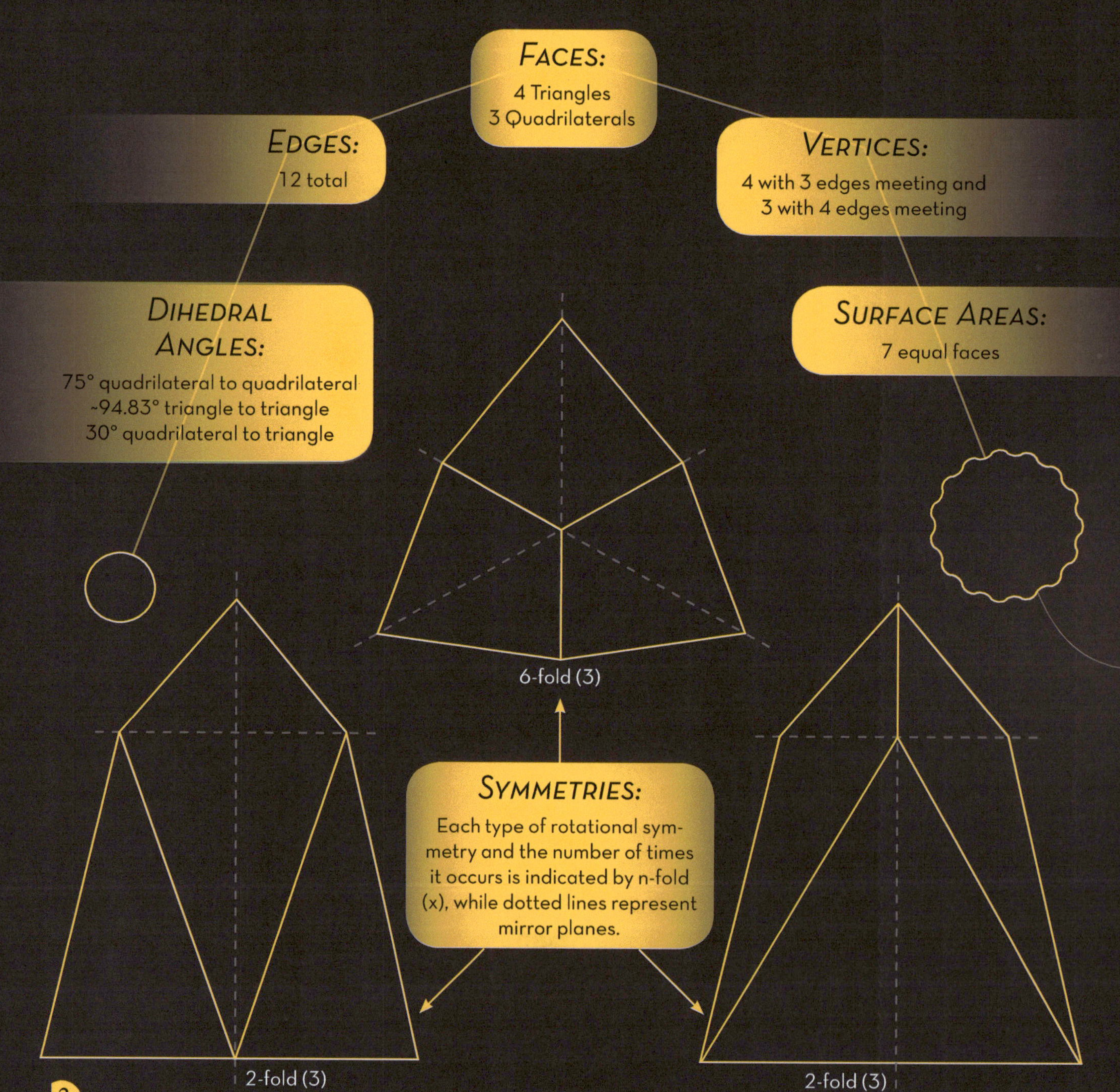

Frank's work helps crack open new avenues of thinking, new ways of understanding our world, and provides insights that help us better understand ourselves. What he is doing, in a nutshell, is exploring the relationship between geometric form and the dynamic principles that underlie the operations of natural and physiological processes, specifically those related to the human being. He did not set out with this goal in mind, but rather followed the trail of a simple question concerning the possibility of a seven-sided geometric form with faces of equal surface area.

As a sculptor, Frank didn't just think about this idea, he made it – "it" in this case being a new geometric form that had never before been discovered. The significance of this is not just that such a form exists, but that the form has meaningful relationships to other phenomenon which are unveiled through a particular way of engaging with the research process itself. In other words, the significance of Frank's achievements lie equally in the realm of the process by which his discoveries were made as in the discoveries themselves.

Frank's work is mold-breaking in a very important way, and this is in terms of the process by which he does his work. The standard methods by which science builds upon itself have some significant blind spots and assumptions that have historically limited its potential (and created some nasty problems along the way, even while helping elsewhere). Most significantly, the methodologies used are generally designed to eliminate what is innately human from the research process, in an attempt to 'control' the situation so that results fit within the assumptions of the experimental design. This style of research provides a certain kind of answer: answers which are geared towards application through reductionistic analysis and control of component parts. Such approaches, and the answers they provide, are proving to be less and less able to provide fruitful metaphors for a sustainable human future.

Frank's work exemplifies an approach that helps bridge this entrenched gap, bringing forth a soulfully engaged style of research that weaves between art and science, allowing the strengths of each to fructify the weaknesses of the other. The result is an integrated style of research that can comfortably deal with both the 'outer' and the 'inner,' without over-privileging one or disparaging the other. In addition to this, his work is mold-breaking in that it constitutes a modern extension of the tradition of sacred geometry, providing a number of well-grounded insights that testify to the integration between humans and the laws of the cosmos.

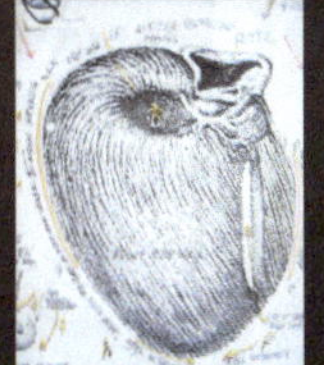
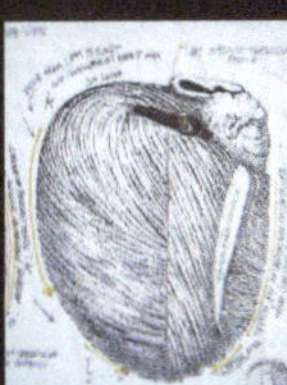
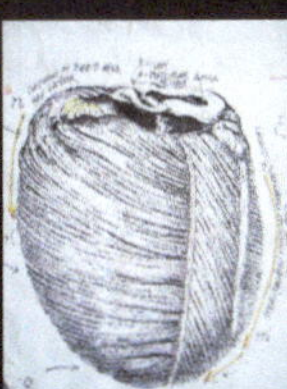
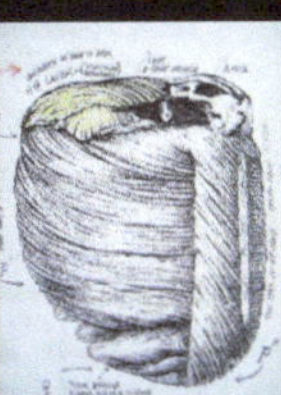
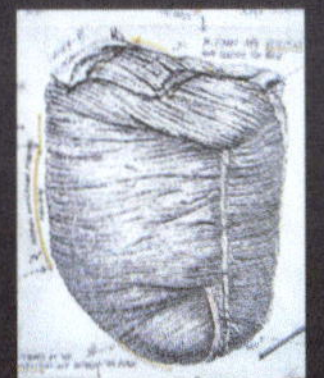
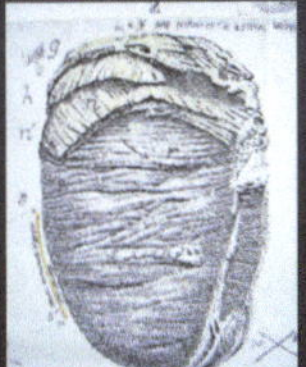
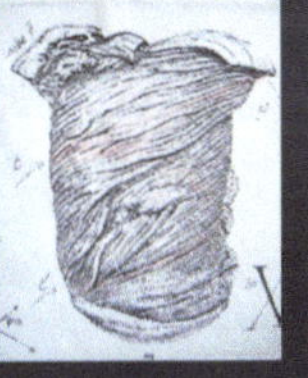
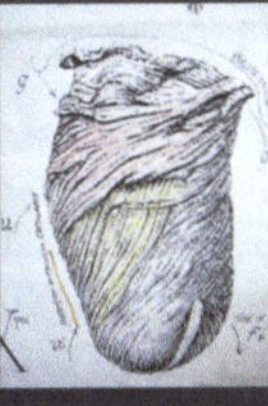

Layers of the heart, Dr. J. Bell Pettigrew

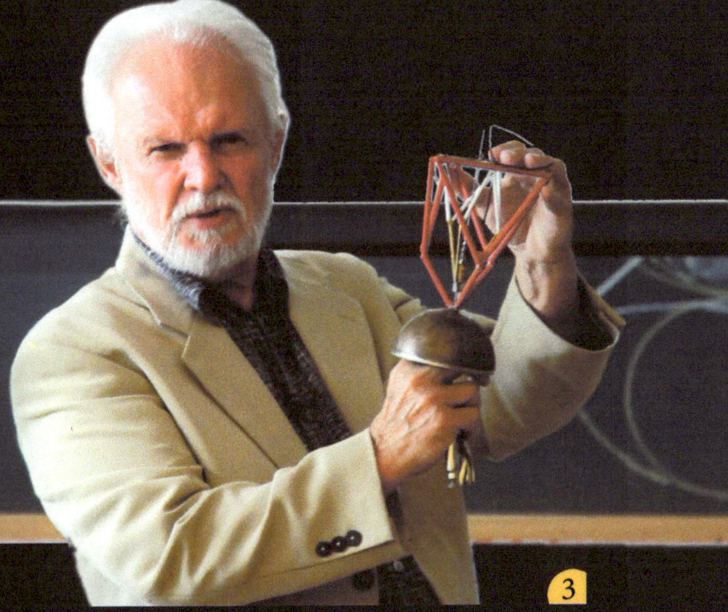

The Chestahedron as an Archetype of Form

POLYHEDRON						
NAME	FACES	POINTS	EDGES	TOTAL	POLYGON	AREA
OCTAHEDRON	8	6	12	28	△	1.0
CHESTAHEDRON	7	7	12	28	△□	1.0
HEXAHEDRON	6	8	12	28	□	1.0

A Platonic solid is a regular, convex polyhedron. This means that it is created by repeating a single shape, like a triangle or pentagon, so that it completely encloses a volume. There are only five Platonic solids: the tetrahedron (4 faces), cube (6 faces), octahedron (8 faces), dodecahedron (12 faces) and icosahedron (20 faces). Each Platonic solid can be transformed into another Platonic solid by pushing its points into planes (truncating them). The cube and octahedron are duals of each other in this way, as are the dodecahedron and icosahedron. The tetrahedron is its own dual. The Chestahedron is similar to the Platonic solids because its faces all have an equal area. It is different in that there are two types of faces, an equilateral triangle and a kite shape (a quadrilateral).

Frank has discovered that the Chestahedron relates directly and exactly to each of the Platonic solids. He has also realized some unique properties of the Chestahedron that require us to think differently about the nature of the Platonic solids in general and how they are formed. He has discovered that the Platonic forms can be created through a phenomenologically based serial transformation beginning with the tetrahedron, rather than requiring them to be paired only with their duals. The serial transformation occurs through the principle of truncation (contractions) as follows:

A tetrahedron's points are truncated (technically this is known as rectifying the tetrahedron). This yields the octahedron (as a transitional form halfway between the tetrahedron and its dual, which is itself). When the octahedron's points are truncated, its dual, the cube, arises. Truncation of the cube only yields the octahedron again, but Frank has found that the cube can transform into the dodecahedron if, instead of pushing points into planes, the *edges* of the cube are pushed into planes. Finally the dodecahedron's points are truncated to yield the icosahedron. The order of the Platonic solids can thus be:

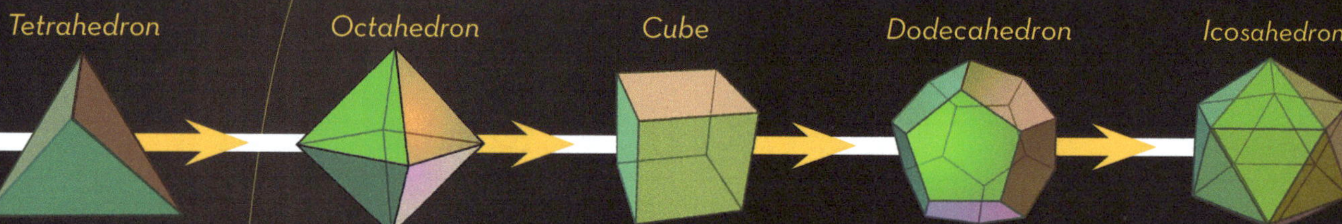

What is important about this sequence is that it uses the contractive principle of truncation to phenomenologically yield a coherent and complete transformative sequence which includes all of the Platonic solids. This is a new way of working with these forms and their order. In both Western and Eastern traditions, different orderings were used, but Frank's is the first sequence based on *phenomenological*, not abstract or concept-based, transformations of the forms.

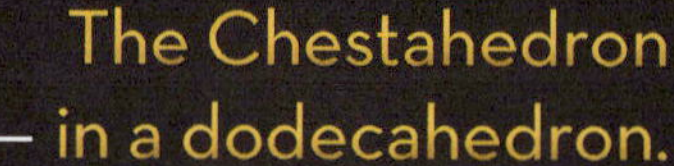
The Chestahedron in a dodecahedron.

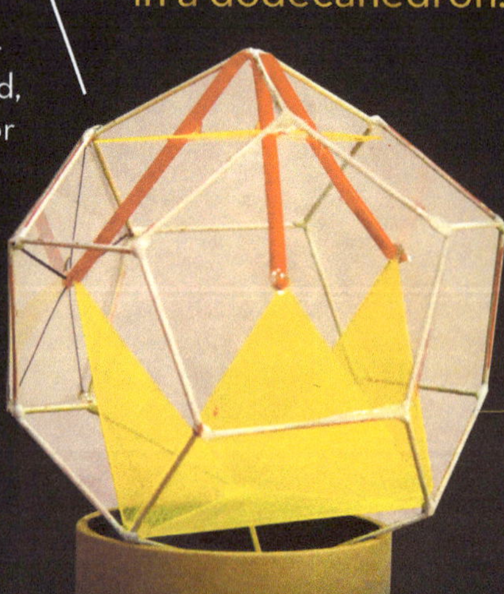

A cube transformed into a dodecahedron by pushing its edges into planes.

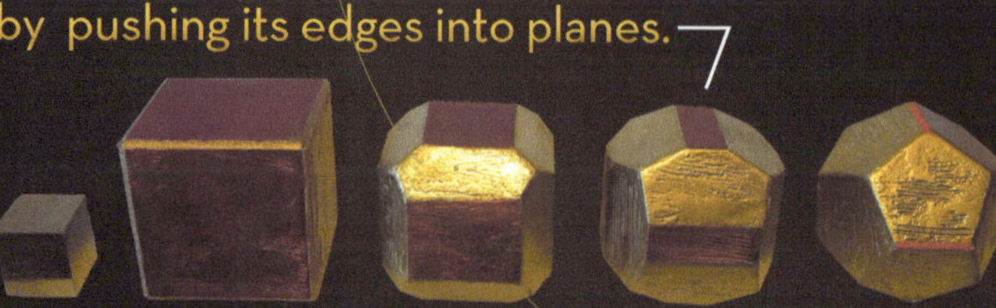

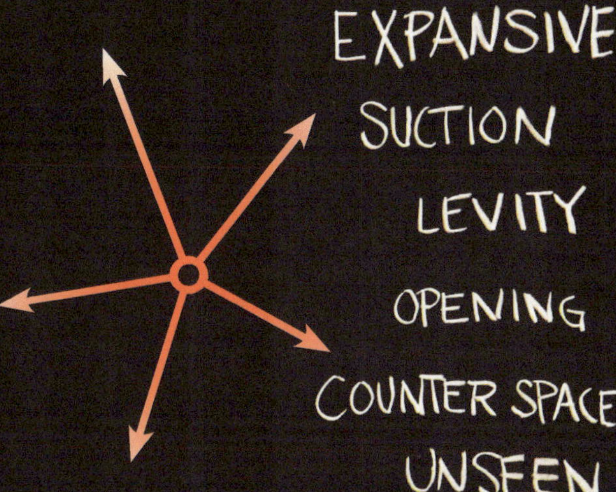

How does this contractive sequencing of the Platonic solids relate to the Chestahedron? We can begin to address this question by noticing that the contractive sequence is only one side of a larger complementarity that is balanced by expansive movements (see chart above).

The Chestahedron is unique in that it can be created from a tetrahedron in *two* complementary ways: one contractive and one expansive. The contractive method works by taking the tetrahedron and spinning it (like a vortex) within a cube (see image at the bottom of page 10).

This procedure has real symbolic significance, as the Platonic solids have been associated with the four elements (thought to be principles underlying the creation of the material world) since antiquity. The tetrahedron is Fire, the octahedron is Air, the icosahedron is Water, the cube is Earth, and the dodecahedron is the Universe, Consciousness, or simply the 5th element. The tetrahedron was identified by Plato in his *Timaeus* as the most fundamental of the forms, in part because it is the simplest, having the least volume to surface area ratio. The contractive mode of creating the Chestahedron requires putting the Fire in the Earth, and transforming the Fire through a vortexial, spiral motion. This is exactly the same pattern identified as the archetype of alchemical transformation. When we understand the symbolic nature of Fire and Earth as will and body, process and product, activity and rest, we can see how the Chestahedron comes into existence as a balancing between these two complementary poles through their unification in a single transformative movement.

But the Chestahedron can also be created in another way, by taking the tetrahedron and opening it up like a flower. This expansive process yields the Chestahedron as an intermediate form between the tetrahedron and another tetrahedron that has sides of exactly double the original.

Frank has discovered that by keeping the dihedral angle (the angle made by the meeting of two faces) constant through the unfolding process, a series of new forms are created (see image at left) that require the faces to become concave, as if a force of suction inside the form were drawing the faces towards its center. Each of these new forms is still based on a seven-sided form just like the Chestahedron, made of three kite shapes and four equilateral triangles, but the kite-shaped faces become concave.

The significance of this process is that it has allowed Frank to see how the seven-sided form is an archetype that appears in all the other Platonic forms. In other words, *all the Platonic forms are unified by the way they relate to the archetype of the seven-sided form*. This expansive process offers a balance to the contractive process: together they paint a more comprehensive picture of the creation of form *in general* through the balancing relations that occur across the boundary between the two polarities. In other words, the task, even in geometry, is to harmonize the contractive and expansive, the material and spiritual, the seen and the unseen. The archetype of the seven-sided form relates directly to both the contractive and expansive poles of creation as a mediator between them. But what is most significant is that through its morphology, it yields a *particular* form: the Chestahedron. The Chestahedron is a unique seven-sided form that has faces of equal area. It is therefore THE form that most represents a harmonic balancing between the material and spiritual, between the contractive and expansive processes of creation.

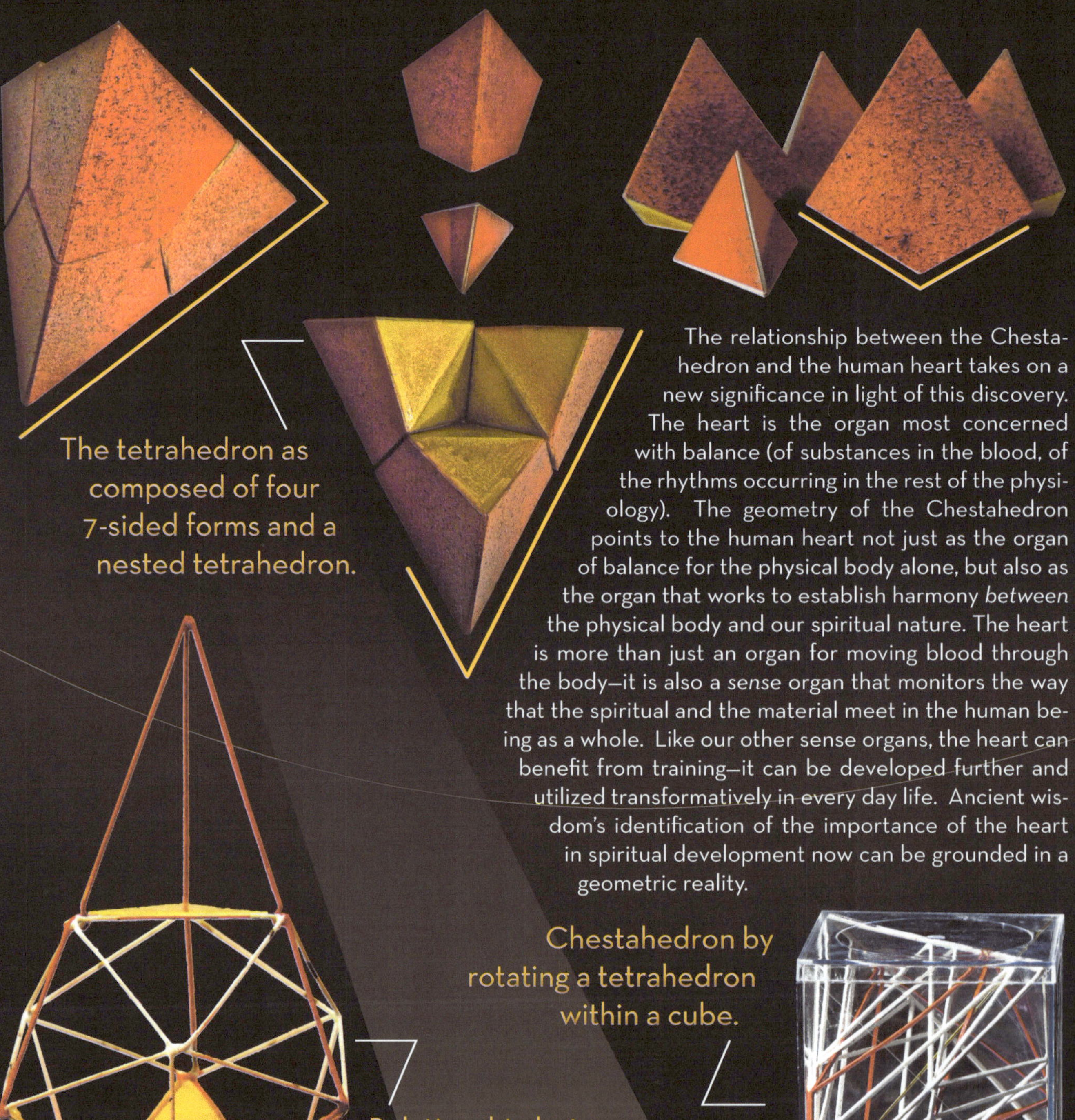

The tetrahedron as composed of four 7-sided forms and a nested tetrahedron.

The relationship between the Chestahedron and the human heart takes on a new significance in light of this discovery. The heart is the organ most concerned with balance (of substances in the blood, of the rhythms occurring in the rest of the physiology). The geometry of the Chestahedron points to the human heart not just as the organ of balance for the physical body alone, but also as the organ that works to establish harmony *between* the physical body and our spiritual nature. The heart is more than just an organ for moving blood through the body—it is also a *sense* organ that monitors the way that the spiritual and the material meet in the human being as a whole. Like our other sense organs, the heart can benefit from training—it can be developed further and utilized transformatively in every day life. Ancient wisdom's identification of the importance of the heart in spiritual development now can be grounded in a geometric reality.

Chestahedron by rotating a tetrahedron within a cube.

Relationship between the 7-sided form and the icosahedron.

The Saturn Seal

The original inspiration behind Frank's search for the seven-sided form came from Steiner's seven planetary seals, each of which has a seven-fold symmetry. Steiner carved relief versions of the seals into the capitals of the large columns in the first Goetheanum, but Frank wanted to know if it was possible to create a fully three-dimensional seven-sided version, rather than an extruded 2-D form.

Frank soon discovered the Chestahedron, but it was only much later that he realized that the bell shape made by spinning the Chestahedron related directly to Steiner's Saturn seal. Amazingly, the exact curve of the bell is implicit in the Saturn seal (see image at top left). Indeed, the entire radiating form at the center of the seal can be thought of as consisting of seven of the Chestahedron bells (see image below and upper right).

Frank had originally called the Chestahedron the "Saturn Form," because it arose as a three-dimensionalization of Steiner's Saturn capital. The planetary seals, although individually still, are meant to morph into each other. The key is to put them into movement through *time*, keeping with the principle of morphological evolution. It is perhaps no simple coincidence that when the Chestahedron is moved through time with a spinning motion it traces out a bell shape whose outline delineates the major features of the Saturn seal, to which the Chestahedron can rightfully be said to have evolved from in the first place. This closes the evolutionary spiral, from the Saturn seal to the Chestahedron, to the Chestahedron bell, and finally back to the Saturn seal.

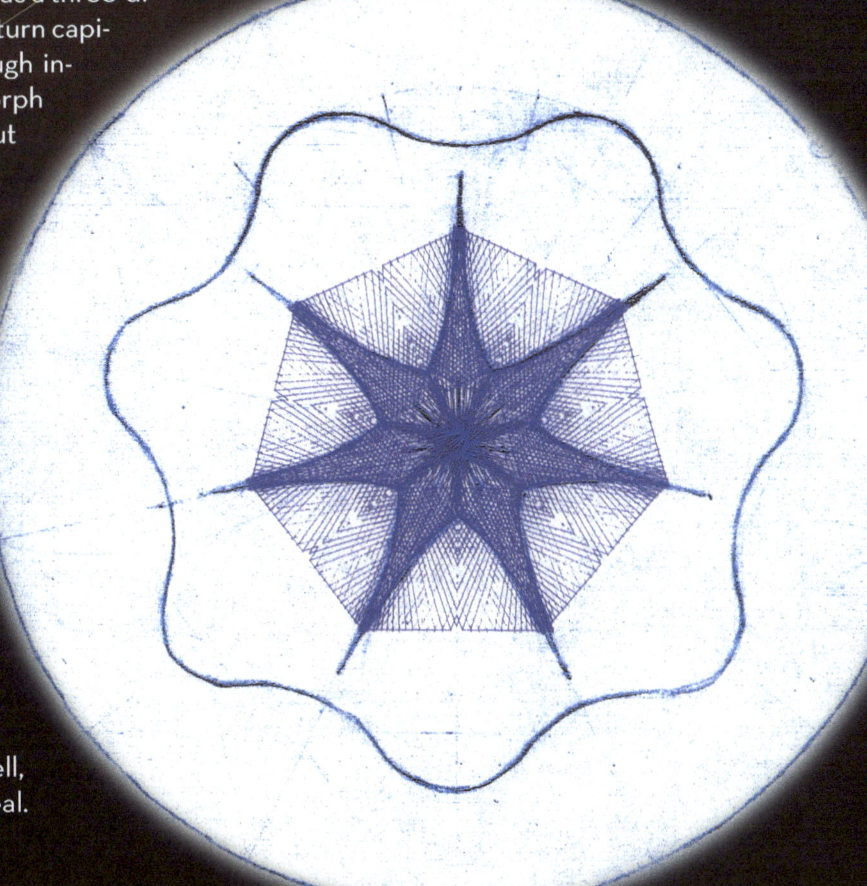

Steiner's Saturn Seal with seven embedded Chestahedra

Discovering the Chestahedron

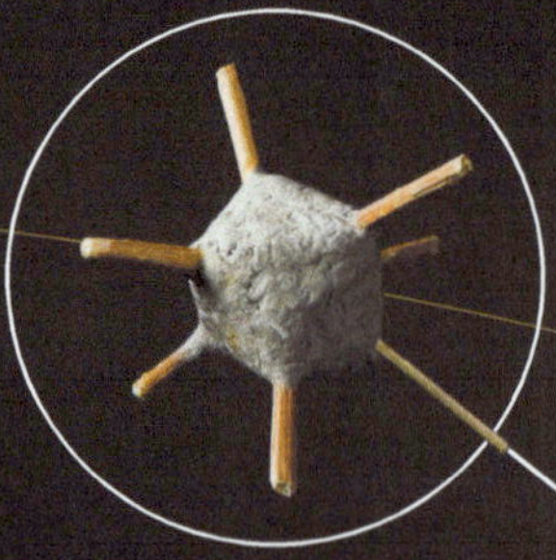

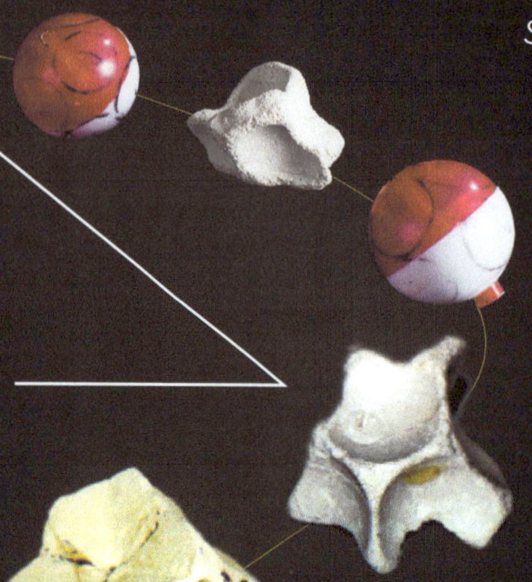

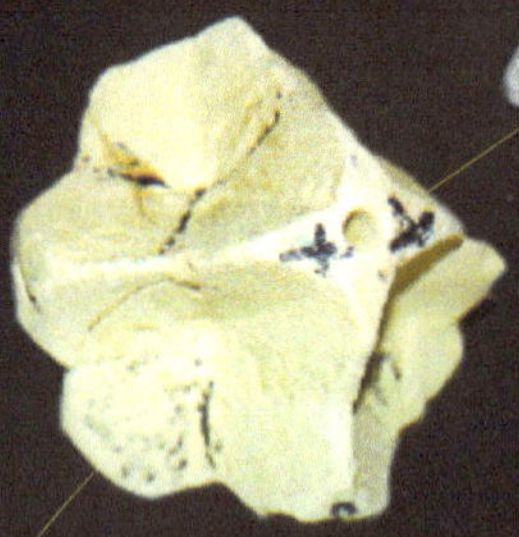

In 2001, just after Frank discovered the Chestahedron, I sat down with him in his small San Francisco apartment to interview him about his discovery.

Seth: When did you begin to work with the Platonic forms?

Frank: When I took Goethean Studies [a program at Rudolf Steiner College], in the Platonic forms section that Patricia Dickson taught. That was my first introduction to Platonic forms. I had never seen them before.

How long ago was that?

One and a half years ago [2000].

And how is the form you work with related to Platonic forms?

Well, there are five Platonic forms and they have equal angles and equal surfaces. The thing about mine is that the surfaces are equal. The Saturn form [the Chestahedron] is a semi-polyhedral form, in that all of the polygons aren't the same. There are two different kinds of polygons: four equilateral triangles and three quadrilaterals, which makes seven. The thing is that there hasn't been a seven-sided polyhedron that has equal surfaces in any of the research. There is nothing

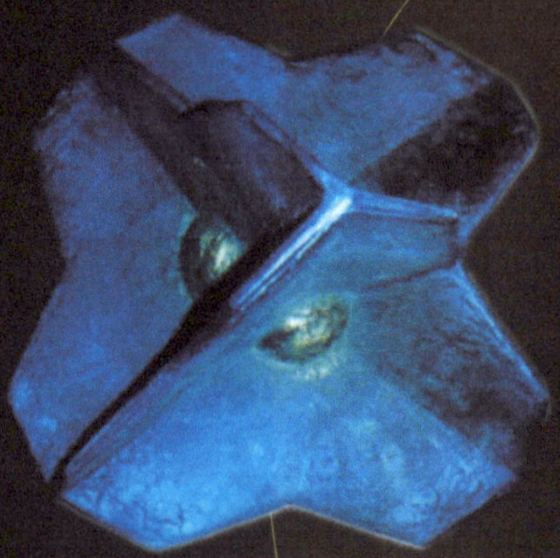

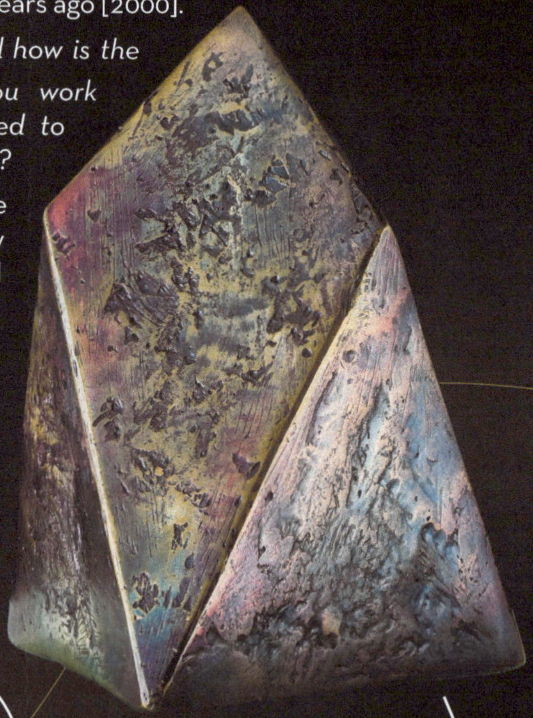

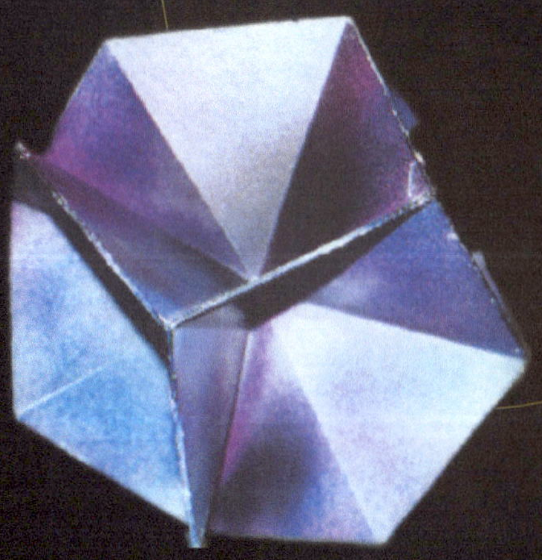

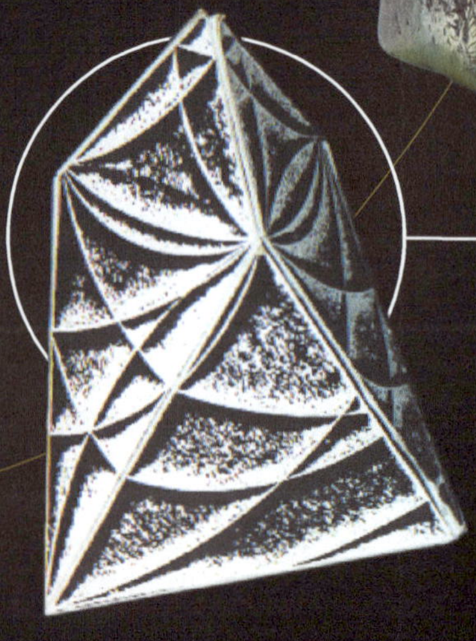

It started with seven sticks in the mud...

The Chestahedron

in the books, in planar geometry, solid geometry, in the artistic or in the science literature. There has never been a seven-sided object with equal surfaces. And that is what is significant about this.

Is that what you set out to do?

No, no. After seeing the seals and capitals at the Goetheanum in 1997, I was inspired. I was told they were powerful forms; they were spiritual forms, which really interested me. But no matter where I looked in the books or in the Goetheanum, I never found any three-dimensional sculptures [of a seven-sided form]. They were always either one-dimensional or two-dimensional. I thought, "gee since I have this sculptural background I have to make one that is three-dimensional." And I didn't do it for two years. I wanted to do it but I didn't have time. I was teaching college, I was busy, or whatever. When I took Goethean Studies and the Platonic forms were taught, I thought "This is my time to do this." I started just before Christmas. I wanted to make a seven-sided object. Everything I did failed; I couldn't do it. I did it with measurements. I tried to do it scientifically. I measured them. I took the seven-sided form in two dimensions and tried to put it in three dimensions and it didn't work I tried to take seven sticks of the same length and put them in a piece of clay and measure them. It didn't work. I tried to take seven balls and push them together with a piece of clay in the center, that didn't work. A whole bunch of things that I tired–forget it, they didn't work.

The Venus form, an inversion of the Chestahedron traced through time.

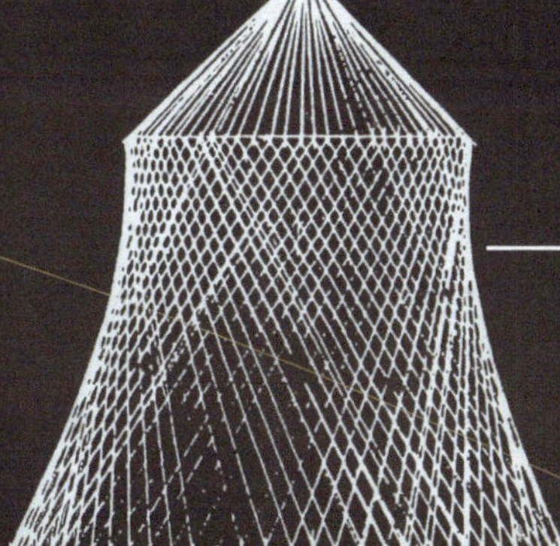

This special bell-shape is made by spinning the Chestahedron.

So then I decided to approach it artistically, which is the way I was trained. When I did I came up with a seven-sided form and I did about three or four of them and I finally refined it until it was quite nice. I showed that to Dennis Klocek [the Director and originator of the Goethean Studies program] and Patricia Dickson and they were really surprised at the shape. So Dennis said something to me that probably started the whole thing. He said to me, "Can you make that objective?" Then I said, "Well what do you mean objective?" He answered, "That's subjective." And what that means is that it is artistic. But can I take it back into science? Can I take it from art to science? A month or so later I came up with the seven-sided form. It came out way beyond my expectations.

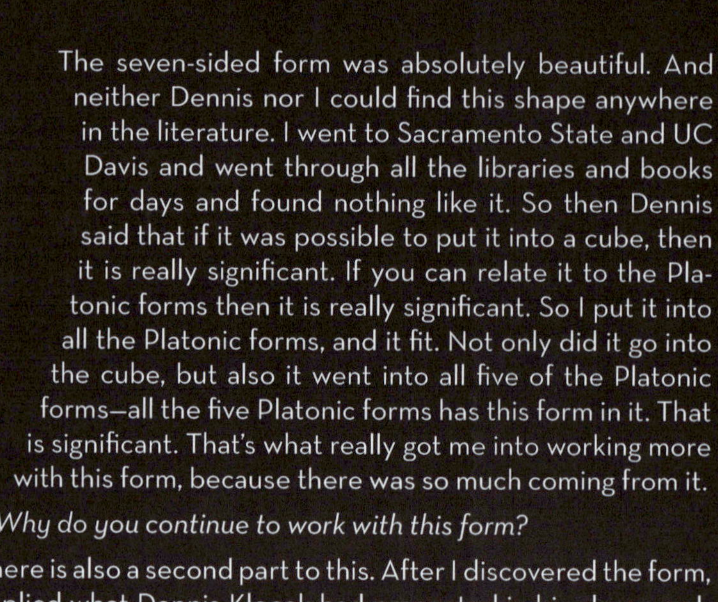

The seven-sided form was absolutely beautiful. And neither Dennis nor I could find this shape anywhere in the literature. I went to Sacramento State and UC Davis and went through all the libraries and books for days and found nothing like it. So then Dennis said that if it was possible to put it into a cube, then it is really significant. If you can relate it to the Platonic forms then it is really significant. So I put it into all the Platonic forms, and it fit. Not only did it go into the cube, but also it went into all five of the Platonic forms—all the five Platonic forms has this form in it. That is significant. That's what really got me into working more with this form, because there was so much coming from it.

Why do you continue to work with this form?

There is also a second part to this. After I discovered the form, I applied what Dennis Klocek had presented in his class on alchemy, about the alchemical cycle of the elements, Earth, Water, Air, and Fire. I could have stopped with the seven-sided form and said, "this is beautiful and here it is," but because of listening to his lecturing I know that alchemy applies to both inner work as well as outer work. So I decided to take this alchemy and apply it to sculpture. So I took the process of the alchemical cycle and applied it to the form, and it just expanded the form into other beautiful forms that I am still working with today. To this very day I am still working with alchemy.

Are you aware of how your work with this form has affected your inner and outer life?

Yes. As far as the sculptural form itself I have developed the capacity to see forms without drawing them–nine sided forms,

The Venus and Bell together in a chalice.

eleven sided forms, thirteen sided forms. I can see them develop in my mind. I can see them without having to actually make them, whereas before I struggled and struggled and struggled to get to that. Now I don't even need to do that. I can see how these forms come, and I make them. So the process has been really shortened. Also because of seeing and working with the alchemical process with my sculptures I also see these processes in my life. If I have a problem in my life, or a question in my life, or I don't understand something in my life, I can take that 'not understanding' and put it through the same process I put sculptures through and I get really clear images of what to do and how to deal with situations. I can apply the alchemical processes that I learned through outer forms to my inner work.

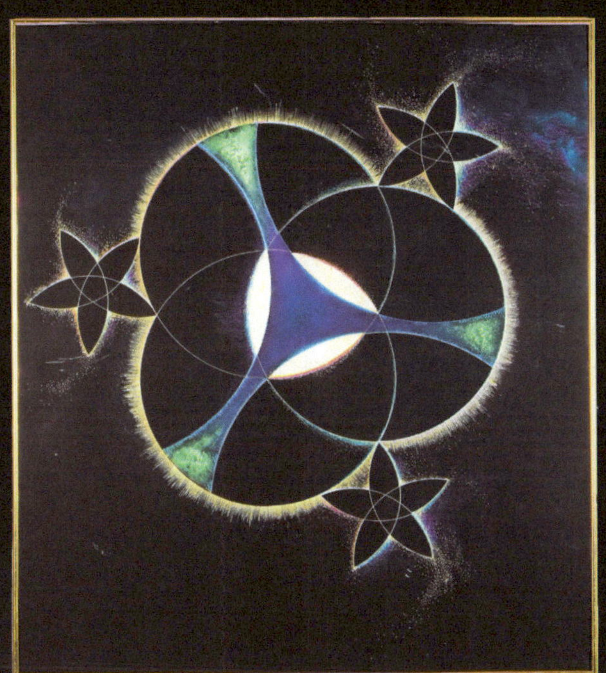

Three interpenetrating Vesica Piscis form the underlying geometry for the Chestahedron.

Wonderful!

Yeah, it is wonderful! I cannot believe how wonderful this is. You can actually take some material, learn the processes of material transformation and apply it to your inner work. I cannot believe how thankful I am for Dennis presenting alchemy in his class.

[End of interview]

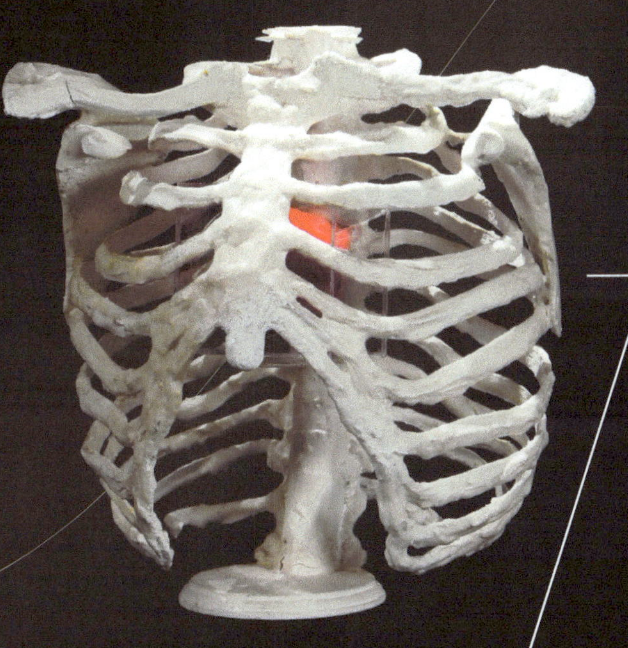

The geometry of the Chestahedron gives a new clue as to why the heart is positioned at a particular angle in the chest.

Tracing dual cones from spun Chestahedra

The Chestahedron can be completely inverted to form a new shape that exactly encloses the Chestahedron.

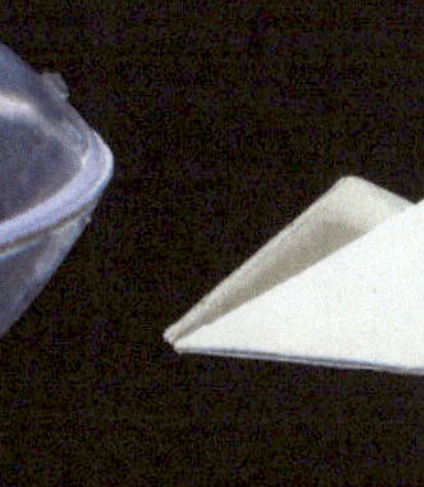

Implications and Applications: The Heart

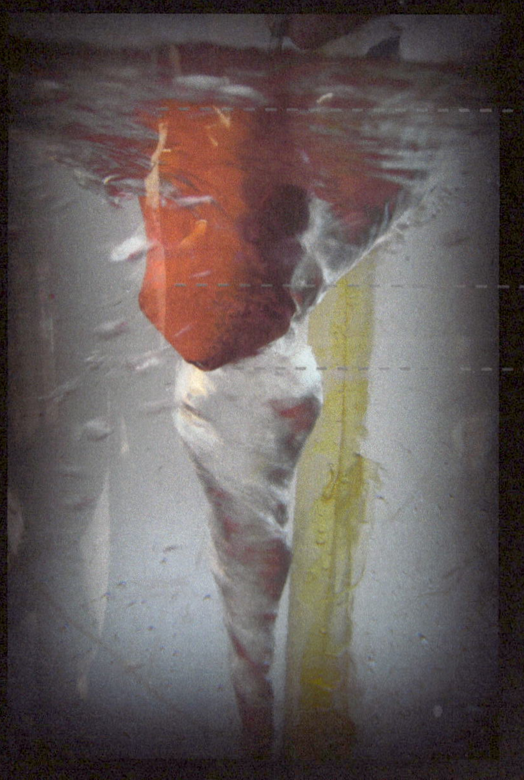

Over a decade has passed since the discovery of the Chestahedron, during which time Frank has delved even more deeply into his geometrical work, discovered a number of tantalizing connections between his forms and worldly phenomena.

Most amazingly, it seems as if the Chestahedron is related to the form and function of the human heart. This discovery, like many of Frank's, seems to have occurred through a concrescence of hard work, experimentation, chance, and perhaps fate. He used a wire-frame model of the Chestahedron dipped in soapy solution to examine the properties of the form in relation to minimal surface areas. The resulting form inspired him to see how the form reacted when spun in water.

When spun vertically, the form produced a normal vortex, but when Frank tilted the form approximately 36° from vertical, something unexpected happened: the water hugged the form tightly on one side but expanded out to form a cavity on the other (see image at left).

Frank, being a sculptor, wanted to know what shape this negative space in the water took, so he used epoxy to fill in the shape layer by layer until the full three-dimensional form was revealed (in blue, above). The connection to the heart came when he realized that a cross-section of the Chestahedron Bell with the new shape attached was exactly like the shape of a cross section of the human heart (right).

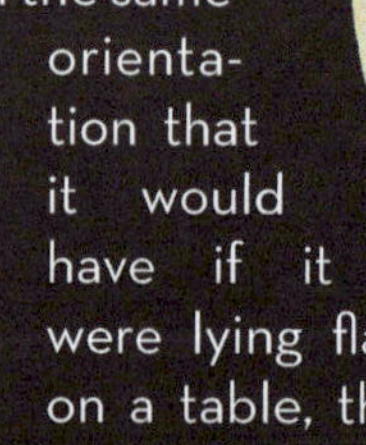

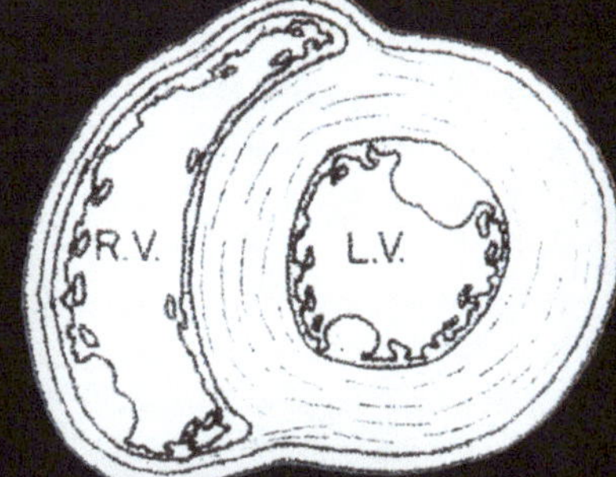

The heart does not sit upright in the chest cavity, but rather is at an angle of approximately 36° from vertical. If one were to take a cube and place it in the chest in the same orientation that it would have if it were lying flat on a table, then the angle of the heart would be almost exactly along the axis from one corner of the cube, through the center, to the opposite corner. This angle in the cube, known as the space diagonal, has a length of the square root of 3 when the length of the cube's side is 1, and sits at an angle of almost 35.3°. When the Chestahedron is placed inside a cube, it fits exactly only when its axis is along the same space diagonal. In other words, the way that he Chestahedron sits in a cube marks the angle at which the heart sits in the chest cavity.

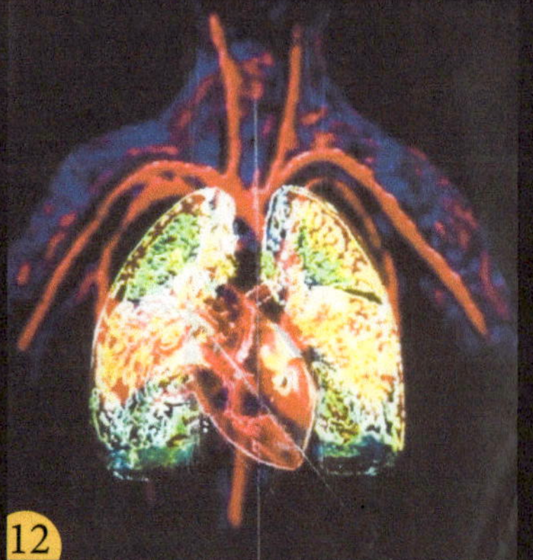

This discovery opened the doorway to many further investigations. The muscle of the heart takes a looping spiral shape, winding around itself in opposing directions multiple times, continuously changing its angle as it does so (see Dr. Pettigrew's drawings of the layers of the heart from the beginning of the book). When the wire-frame Chestahedron is spun, the shape, traced out through time, results in the Chestahedron Bell (right). Three cones are formed in this process. One cone fills the interior space, one hugs the outside, and a last cone is made from the projection of the tip of the bell (left).

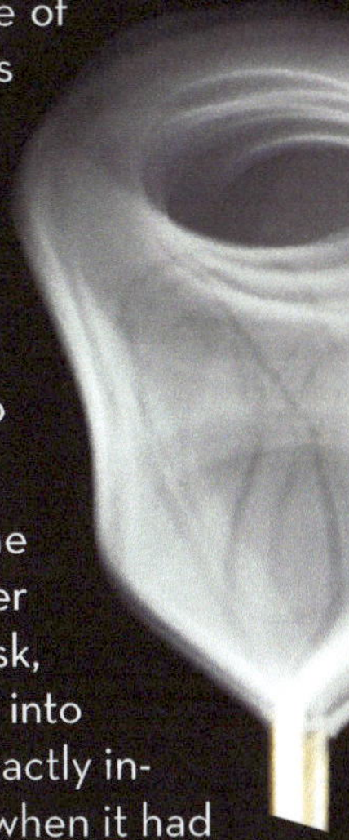

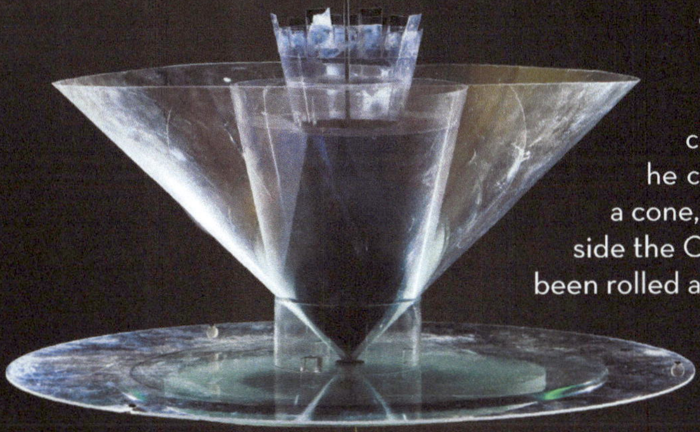

Frank found that if he took a piece of paper cut into a flat circular disk, he could roll it up on itself into a cone, and that it would fit exactly inside the Chestahedron's center when it had been rolled around itself four times. The angle of the cone was 26°. Moreover, it turns out that the cone that hugs the outside has exactly the same angular measure of 26°. This means that if a single extended cone of 26° is cut in half laterally, the tip will fit exactly in the center of the Chestahedron while the Chestahedron as a whole will sit exactly in the other half. The significance of this took Frank some time to decode.

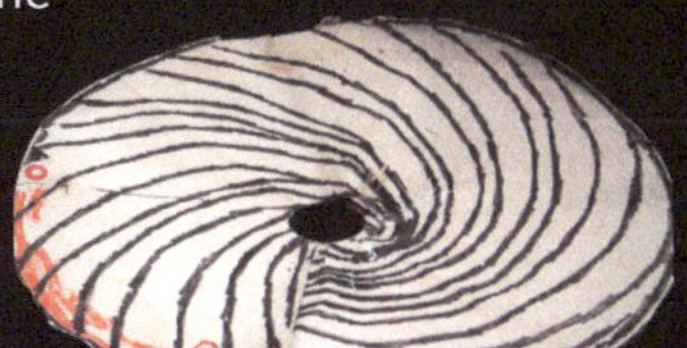

When lines are drawn on the circular disk, and it is rolled up on itself into a cone, the lines mirror the complex shifting layering of the fibers of the heart muscle. When Frank tried this with a single disk he only got four layers. But he found that if *two* flat circular disks are put on top of each other, cut along radii that are then connected to each other to form a single continuous spiral, then when rolled into a cone of 26° a more complete analogous layering occurred. If the disk is rolled seven times, there is exactly enough of a partial disk left to roll in the opposite direction a single time. This produces a strong seven-layered circle and a flimsy wall of only one layer, which can easily be flattened with a hand around the seven-layered circle, producing a form that almost exactly mirrors the way the thinner muscle of the right ventricle of the heart cups the thick-walled left ventricle in a crescent shape. This arrangement is significant, because it matches the delicate seven-fold layering of the heart found by Dr. Pettigrew's careful dissection technique.

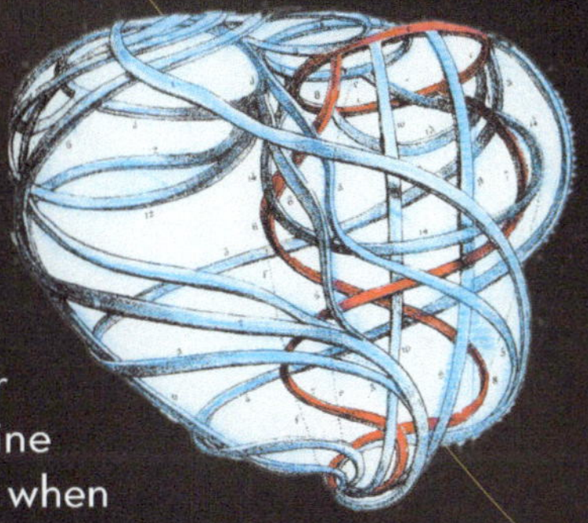

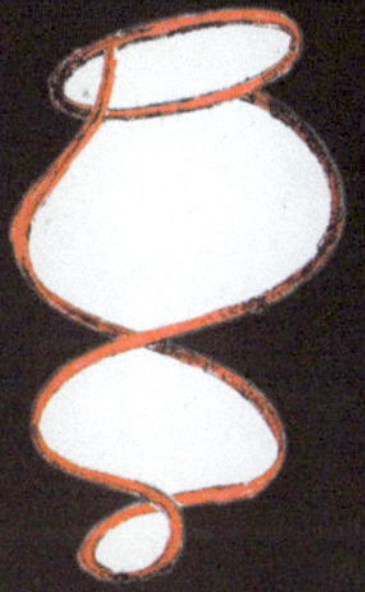

Importantly, some of the muscle fibers spin around the thin apex at its lower end before looping back up (right). A particular straight line (a chord), when drawn on the circular disks, closely approximates the complex looping when the disks are rolled up. This particular straight line is a chord that cuts the circle exactly halfway between its center and its circumference. This line is thus the exact same line that is formed when two circles meet in the Vesica Piscis. If the length of the radius of the circle is 1, then the length of the line is the square root of three.

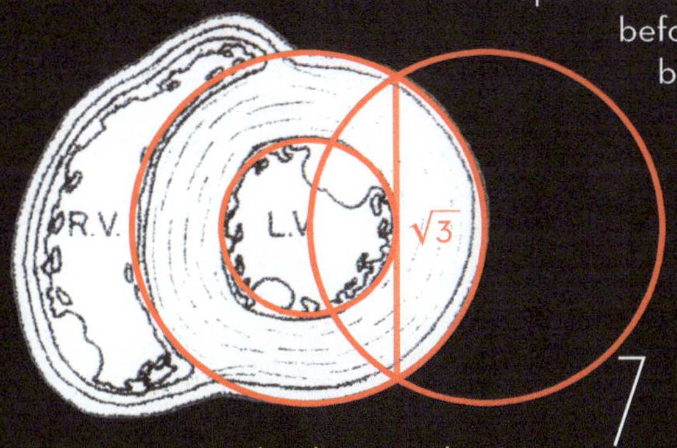

R.V.
L.V.
√3

Root 3 joins the heart, the Vesica Piscis, and the Chestahedron.

Astonishingly, in what seems almost an unbelievable coincidence, the cross-section of the spun Chestahedron, with its inner and outer cones of 26°, marks not only the thickness of the wall of the left ventricle, but also joins the sacred geometry of the Vesica Piscis and its root-3 intersection in a new way with the spiralling layers of the heart muscle. We see that root-3 plays a role in connecting the Chestahedron to the heart in two completely different ways: once in a static way through the positioning of the heart in the chest cavity, and once in a dynamic way through spinning the Chestahedron to yield the two cones that produce the thickness of the muscles of the left ventricle.

Even further, if we examine the way that the geometry of the Chestahedron itself forms when a tetrahedron is spun inside the bounds of a cube, we see that it only results when the axis of rotation is along the root-3 space diagonal of the cube. The Chestahedron itself is a form born out of transformative movement. If we look more closely a this geometrical birthing process, we see that within the cube the sequence begins with the contraction of the tetrahedron, which then produces the Chestahedron before contracting into an octahedron.

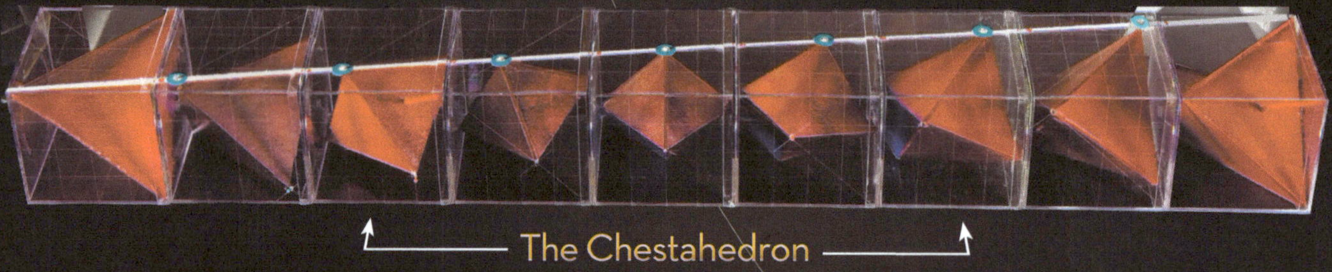

The Chestahedron

The octahedron has been linked for ages to the element of Air, which has the alchemical connotation of reversal. It is just at this point that the contraction reverses, now expanding again, first into another Chestahedron and then finally ending with a last expansion back into a (now rotated) tetrahedron (above).

This process of contraction and expansion should immediately bring to mind the rhythmic beating of the heart. What is more, the ventricles themselves are involved in reversals, changing the direction of the blood. As Craig Holdrege, director of the Nature Institute indicates in his phenomenologically-oriented book *The Dynamic Heart and Circulation*, "the right side of the heart brings vertically flowing blood into the horizontal and the left side of the heart brings horizontally flowing blood into the vertical" (p.10-11). The repeating theme of the heart—it's archetype—seems to point towards integrating diverse tensions through rhythmic oscillations, particularly by the creation and utilization of vortexial forms. Ulrich Morgenthaler, CEO of Fourm3, illuminates this perspective in an article published in Das Goetheanum magazine, No.20-10:

"If the heart were [merely] a pump, the paper-thin tissue at the apex of the left ventricle could never withstand the developing pressure. However, from the perspective of a vortex model of the heart, it becomes understandable why this part of the heart is never exposed to these higher pressure dynamics."

The heart is like a master practitioner of the spiralling art of aikido, meeting force with relaxation and using relaxation to skillfully manipulate force, all with an eye to harmony.

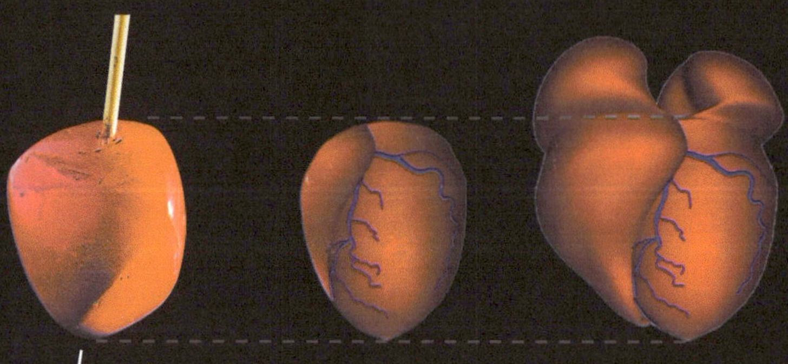

The Chestahedron in minimal surfaces mirrors the shape of the left ventricle of the human heart.

Implications and Applications: The Earth

In a curious 1924 lecture by Rudolf Steiner on volcanism, he spoke of the shape of the Earth in relation to the shape of the tetrahedron, but rather than a regular tetrahedron, he imagined a sort of rounded tetrahedron (see his original chalk drawing below).

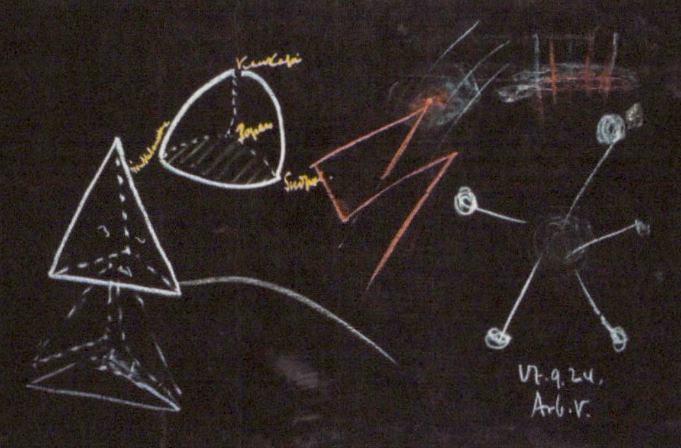

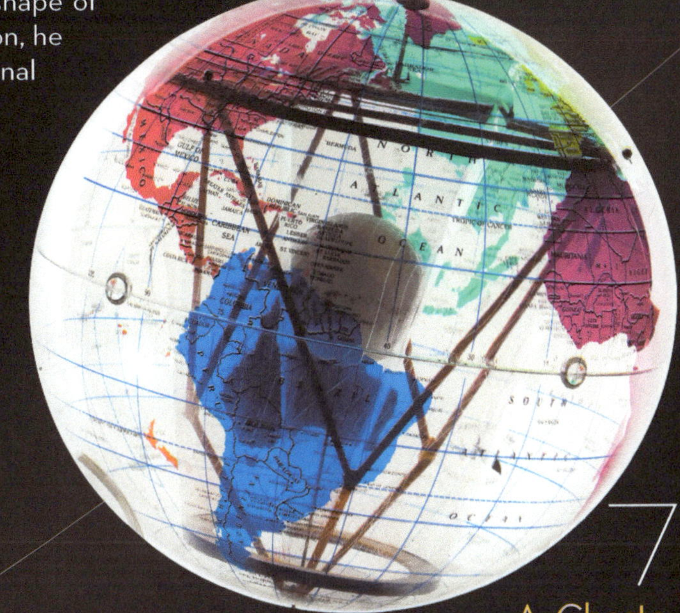

A Chestahedron in the Earth.

Frank, inspired by this lecture, wondered what would happen if he placed the Chestahedron inside a globe. In his lecture, Steiner placed the four points of his "kind of tetrahedron" on the globe at Central America, the Caucasus mountains in Europe, and Japan, with the final point on the South Pole. However, these three points don't form an equilateral triangle. If a point in northern Japan, near Mt. Osore, is taken as an initial point, then the other two points can be found. One is in southern Macedonia, and the third is in the middle of Nebraska. If this equilateral triangle is taken to be the base of a Chestahedron, then it happens to be just the right size so that the Chestahedron's fourth point exactly reaches the South Pole.

This relationship seems a bizarre coincidence—why would the Chestahedron (which is literally formed by opening a tetrahedron at one point, and which thus can rightfully be called a "kind of tetrahedron"), when placed in the Earth according to Steiner's indications—made more geometrically exact—have its fourth point meet the South Pole? That this occurs depends upon the latitude at which the Chestahedron's base triangle is placed. Is it a coincidence that the only latitude at which this relationship holds is the same latitude identified by Steiner?

TETRAHEDRON, SOMETHING OF A TETRAHEDRON AND THE CHESTAHEDRON IN THE SPHERE OF THE EARTH

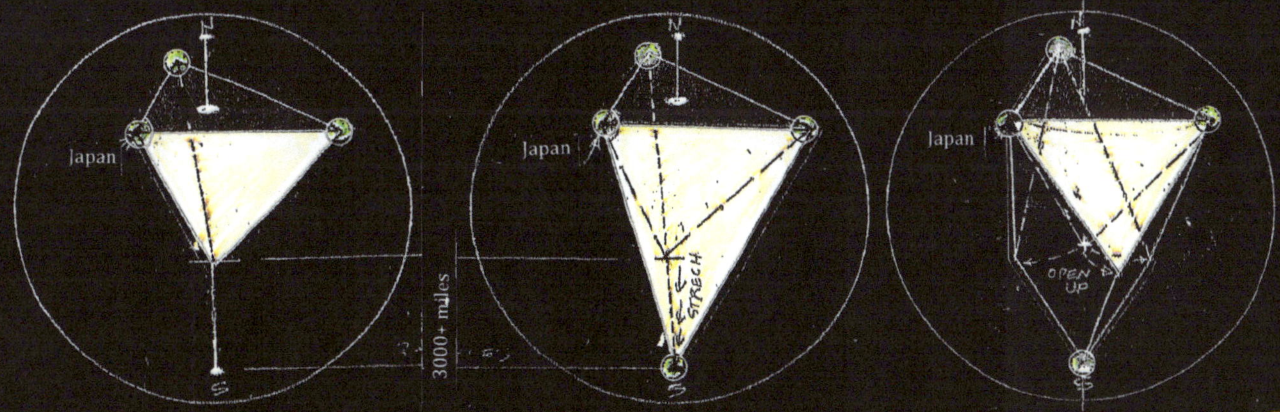

The Caucusus mountains start their northwesterly journey just at this latitude. This mystery only inspired Frank to look into further possible connections between the Chestahedron and the Earth.

As usual with Frank's work, illuminating further connections requires a deeper look at the actual geometry.

There are exactly two hexagons hidden inside the Chestahedron. They are revealed as cross-sections of the form, taken parallel to the base triangle. One is formed by cutting the Chestahedron almost midway up. This hexagon, or alternatively, six-pointed star, is formed at precisely the point where the cross-sectional plane makes equilateral triangles on the faces of the three triangular "petals" of the tetrahedron it intersects (see image at left, with equilateral triangles and upper hexagon in red).

The second hexagon is formed nearer the base triangle, and is formed such that three of its edges lie *inside* the Chestahedron, whereas the upper hexagon's edges are all on the periphery of the Chestahedron (see image at right). These two hexagons are almost exactly the same size (the lower is slightly smaller) and lie directly above one another. The points of the upper hexagon all lie on the *edges* of the Chestahedron, while the points of the lower hexagon all lie on the *planes* of the triangle faces. We can imagine that the upper hexagon encompasses and restricts the Chestahedron from the outside, while the lower hexagon expansively pushes outward from the inside. Again we can see how the Chestahedron is a form that seems to be built entirely out of the activity of harmonizing two tendencies that would normally only be in opposition to each other.

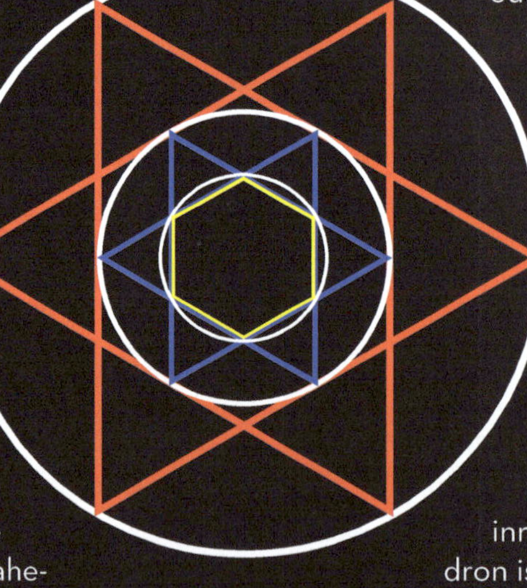

When a six-pointed star (below, in red) is made from the upper hexagon, a circle can be inscribed in its middle. That circle meets the six-pointed star in six new points, which can be the basis of a further six-pointed star (blue) that is exactly half the size of the first. This smaller star has a hexagon in its middle as well (yellow). As any good geometer knows, a hexagon is the 2-dimensional shadow of a 3-dimensional cube standing on its point.

Frank made this cube, which sits right at the center of the Chestahedron, standing on its point. The relationship to the Earth becomes apparent only when a sphere (the inner-most white circle) is placed around the cube and the Chestahedron is sized to fit exactly in the globe of the Earth. When this is done, the size of the sphere is such that it is directly comparable to the size of the inner core of the Earth. Again, root-3 makes an appearance, being the length of the diameter of the sphere in relation to the length of the side of the cube.

We have already met the central cone of 26° that fits exactly into the center of the Chestahedron. If the Chestahedron is placed in the Earth with its tip at the South Pole, and the cone is placed into the Chestahedron, then if a sphere is dropped into the cone it will settle right in the middle of the globe (see image previous page), but only if it is scaled to be the size of the inner core of the Earth. In other words, there are two different ways in which the Chestahedron yields the relative size and position of the inner core of the Earth.

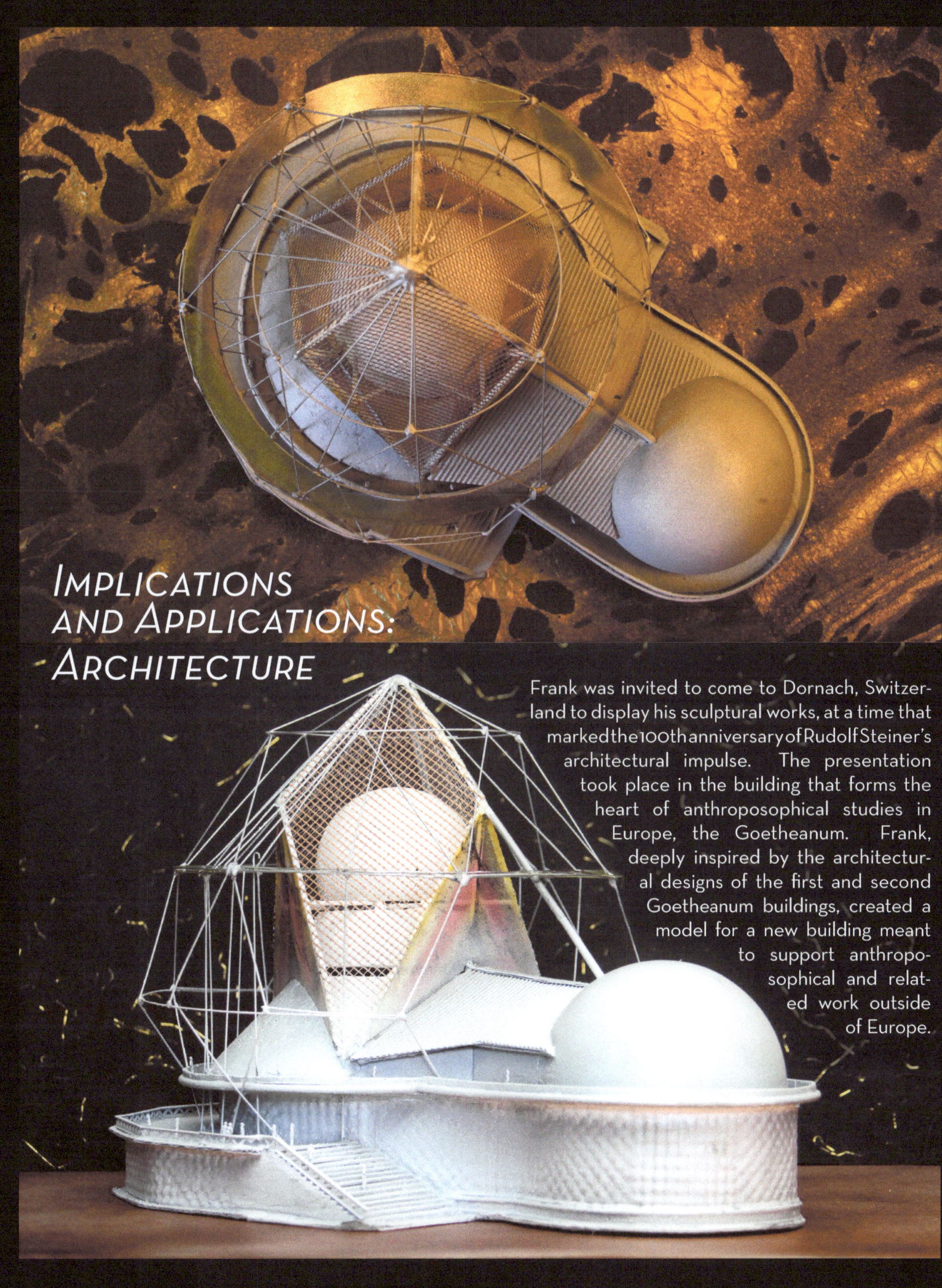

Implications and Applications: Architecture

Frank was invited to come to Dornach, Switzerland to display his sculptural works, at a time that marked the 100th anniversary of Rudolf Steiner's architectural impulse. The presentation took place in the building that forms the heart of anthroposophical studies in Europe, the Goetheanum. Frank, deeply inspired by the architectural designs of the first and second Goetheanum buildings, created a model for a new building meant to support anthroposophical and related work outside of Europe.

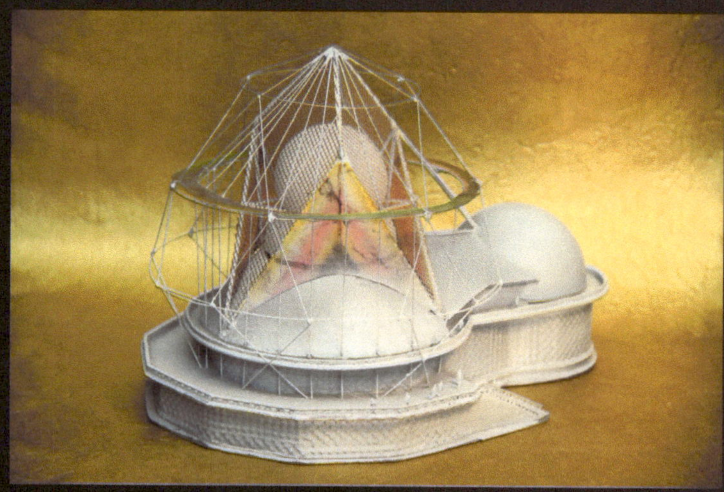

Frank was inspired directly by both buildings, and has taken up the impulse to further the architectural reflection of the continued evolution of anthroposophy in his own design, which is a further metamorphosis of the designs and principles Steiner used in the first and second Goetheanums. The design is centered around the unique geometry of the seven-sided form that he discovered. This form, the Chestahedron, derives its name from its connection to the heart: it is a proposed geometrical template for the form of the human heart. The Chestahedron therefore embodies the principles behind the development of the kind of heart-thinking that is so important for our time. The building is meant to carry forward the necessary outer manifestation of the changing nature of anthroposophy itself, and would provide an appropriate space for lectures, workshops, mystery plays, musicals, eurythmy, visual arts, and meditation. If built, the space is intended to support public outreach and provide a powerful physical representation of the life-giving nature of anthroposophical activities in the world.

Building Features

The geometry of the structure is based on the form of the human heart. Two cupolas meet in the main part of the building, in a way reminiscent of Steiner's ground-breaking design for the first Goetheanum. However, in this case the cupolas interpenetrate vertically and are suspended by the Chestahedron above. The upper cupola is based on a geometry of the number 5, while the lower cupola is based on the number 7. Together they meet in a plane that forms a perfect six-sided hexagon.

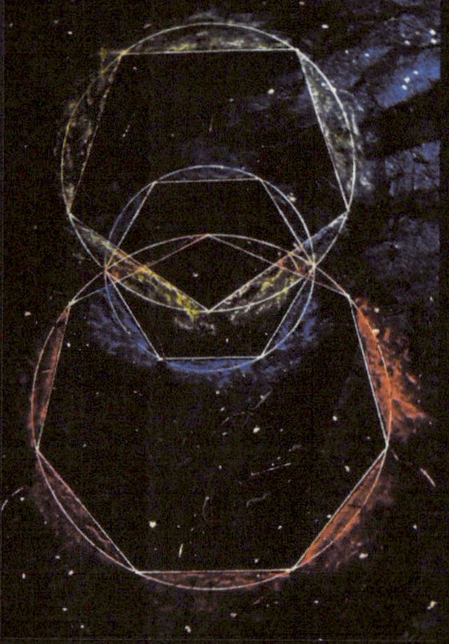

While the main domes are suspended from and sit inside the Chestahedron, the Chestahedron itself is embedded in its dual form, the decatria (dek-a-TRIA: a form with 13 faces). The Chestahedron and decatria together embody the principle of reversal, and the principle of levels of self-embeddedness and self-unfolding that are key recurring patterns in anthroposophical work. Half of the supporting wires from the decatria are structural, while the other half are made of a thick nylon that will allow the strings to literally be played as a musical instrument with an appropriate bow. The building itself thus embodies a musical principle as a part of its very structure, and will "sing" its geometry gently with the blowing of the wind, which will make the strings resonate at their fundamental frequency.

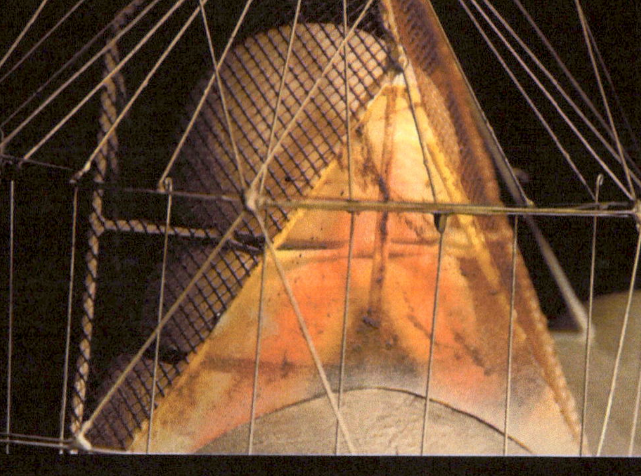

Unlike the capitals in the first Goetheanum, there are no capitals inside the building; the structure is not supported by columns from below, but is rather suspended from above, from the outside. This allows for seven internal, non-weight-bearing columns which are capped by seven individually made bells. The seven bells are based on the transformative sequence of Steiner's seven planetary seals and the geometry of the Chestahedron, and increase in size by a factor of 1/7th. Where the capitals in the first Goetheanum formed a transformative sequence of sculptural form, the bells form a transformative sequence of audible tone. The sequence of tones yields a transformation at a higher octave, and can fill the space in a way that can be shared directly and inwardly that is not possible with purely visual forms.

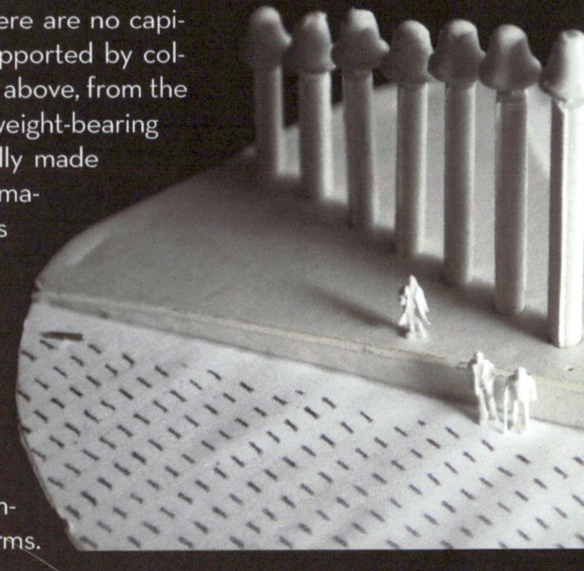

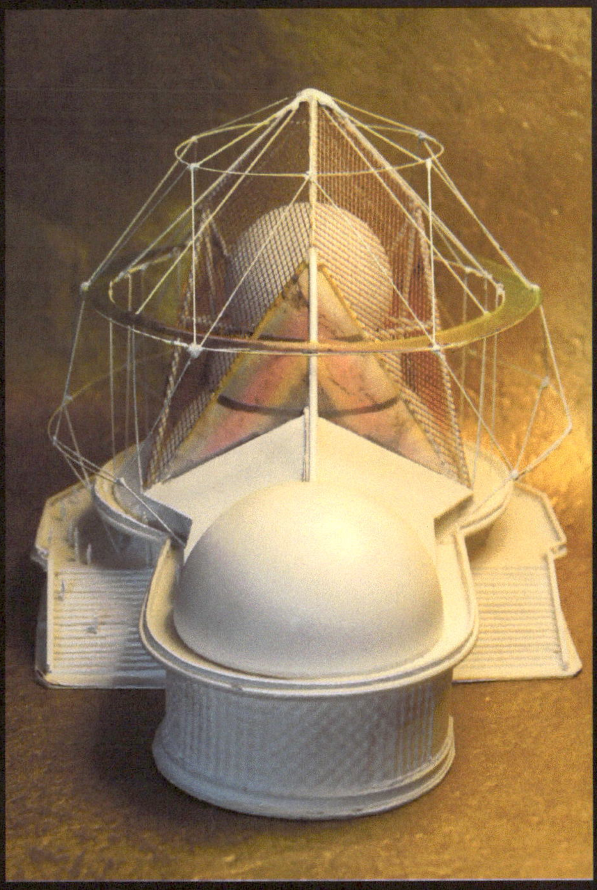

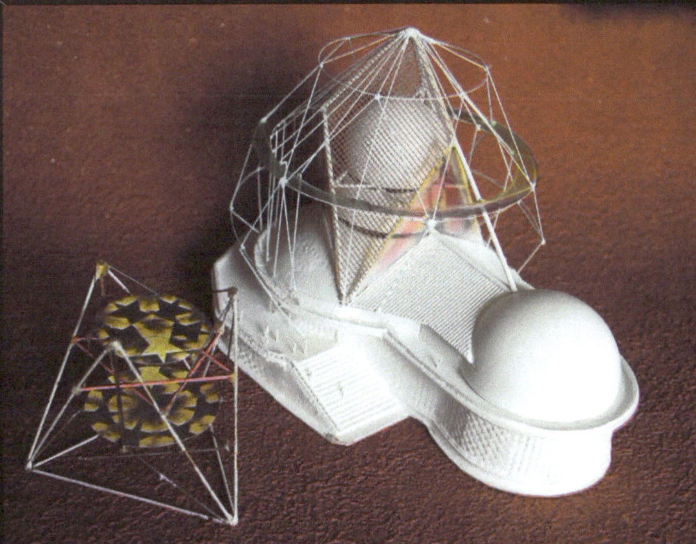

Rather than force the skin of the building to hold only one color, it will instead be coated with a surface that will reflect the color projected onto it by a series of surrounding tourmaline-shaped panels of colored glass. The color of the building will be organically formed out of the seasons and daily weather conditions, and reflects an integration of the natural and spiritual environments as they change throughout the year.

Earth and Moon

Frank has discovered a new way to square the circle (constructing a square and a circle that have equal perimeters). His method uses only a straightedge and compass, and is the first to work from the inside out using the Vesica Piscis.

The mathematics of this construction are known exactly. Measurements of the circumference of the circle and the perimeter of the square agree to within 99.9+%, with the perimeter of the square ever so slightly smaller than the circumference of the circle.

It was proven in 1882 that it is impossible to perfectly square the circle using a finite number of steps with a straightedge and compass. This is a result of the Lindemann-Weierstrass theorem, which was involved in showing that the number pi is transcendental, and thus a non-algebraic number. Straightedge and compass constructions involving non-algebraic numbers can only be approximations.

Within the discipline of sacred geometry, it is well known that when a circle is squared, the relationship between the relative sizes of the Moon and the Earth is made explicit. If the Earth is scaled to fit exactly within the perimeter of the square, then a circle of the same perimeter as the square will pass through the center of the Moon, if it were to touch the Earth. (See the drawing on page 40.)

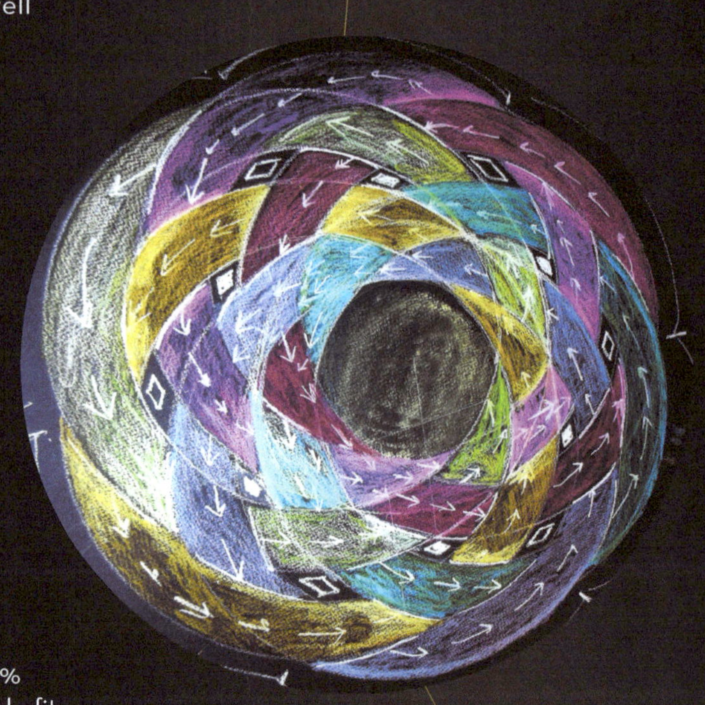

However, Frank's construction method is unique, because it not only yields the Earth-Moon relationship, but also delves *inside* the Earth, revealing a more subtle relationship between the size of the Moon and the size of the *interior* of the Earth.

The Earth has both an inner and outer core; together they form a sphere 6972 km in diameter. The Moon has a diameter of 3474 km, which is almost exactly half that of the Earth's core (~99.7% agreement). In other words, two Moons, side by side fit exactly into the Earth's core. Frank's method yields points (noted by small white circles on the drawing on p. 40) that mark the relative size of the Earth's core, and thus its relationship to the Moon.

Between Art and Science

Rudolf Steiner said that geometry is "a knowledge which appears to be produced by man, but which, nevertheless, has a significance quite independent of him." Frank's work perfectly exemplifies this dynamic interplay between the creative effort of producing entirely new forms and the difficult but rewarding search for their possible significance.

This way of working can be contrasted with the current split between "pure" and applied sciences. The history of mathematics is replete with cases where freely created abstract mathematical relations with no obvious ties to the physical world eventually become central to solving "real-world" problems. But when the link between the pure logic and its potential significance is not held in dynamic balance in the individual, a great opportunity for the two poles of activity to mutually enliven each other is lost.

Frank's style of working serves as a kind of healing of this gap, constantly weaving between the abstract logic of geometry and the way that the relations found there illuminate the world around us. Importantly, the way that this weaving between worlds is accomplished is by virtue of a rhythmic artistic process. This allows new directions to be explored not just on the basis of purely logical consequences on the one hand or purely material considerations on the other, but in a way that is colored by a deep engagement of the soul in the creative activity as it unfolds *between* the abstract and material.

It is a process that is recursively open to both new logical insights and new material discoveries, and acts as a bridge between them. The bridge itself is constructed out of soul activity as it shifts in response to the tension between the logical and material, between the spiritual and physical, and between the archetypal and the particular. As in a suspension bridge, the functionality depends upon harmoniously balancing forces from multiple simultaneous directions, not by eliminating the tension, but by *maintaining* it as the very source from which new creative work springs.

That this description of Frank's process mirrors the previous qualities discussed in relationship to the heart is no accident. The integration of art and science occurs by virtue of the heart's ability to hold the tension of opposites in a way that Steiner called "thinking with the heart." Developing this capacity is what has led Frank to a way of doing artistic science, which can rightfully be called *heartistic science* for this reason.

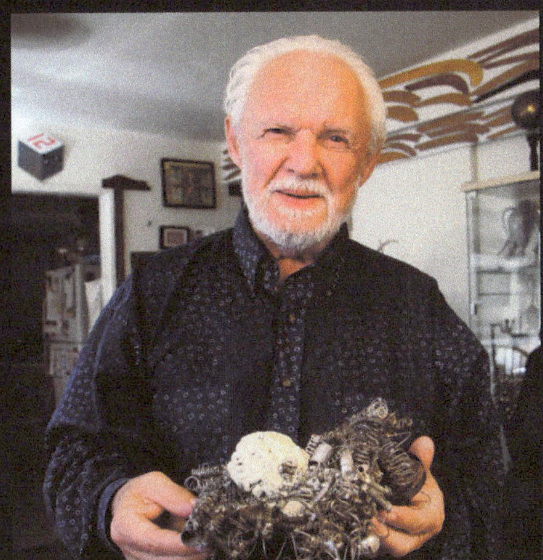

The process has implications for both science and art, and points to the need for modern science, which is still laboring to free itself from the spell of the enlightenment-era notion of "objectivity," to open its doors to a more subtle and human way of working that does not discount what it cannot immediately understand. Similarly, the arts can benefit from opening doors to science. Rather than ignoring or even disparaging science, the arts can engage with science both as a source of inspiration and as a way to bring to art the same kind of process that when directed at purely material relations yields scientific knowledge.

Goethe called for something similar in his "delicate empiricism," which inspired Rudolf Steiner's creation of anthroposophy, or spiritual science, which inspired Frank Chester, who is now carrying the impulse forward in his own unique way.

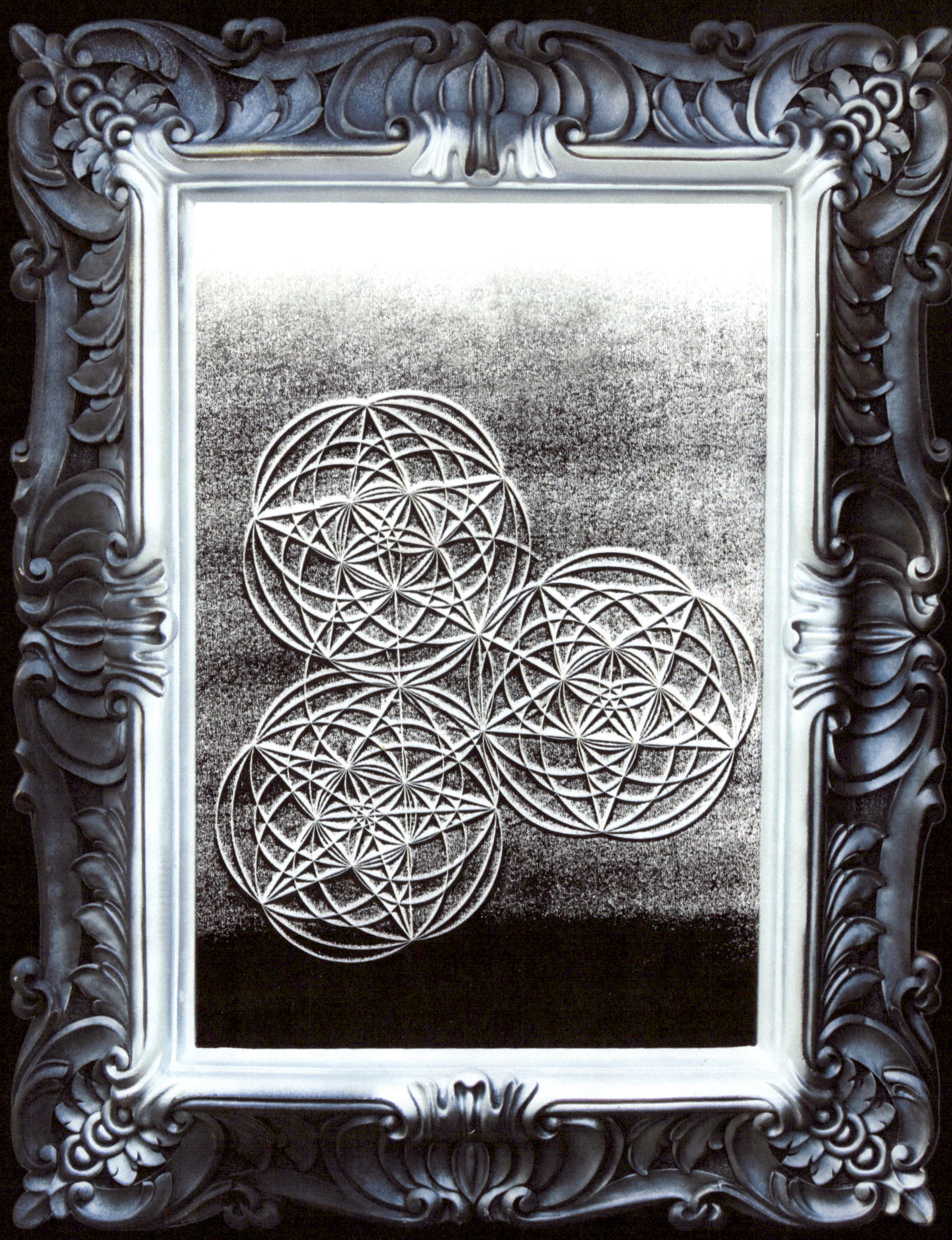

The 2-dimensional geometry
underlying the Chestahedron fully revealed.

Making Connections

The importance of Frank's work is being noted, but deserves much wider attention. The Mathematical Section of the Anthroposophical Society has recognized his lasting contribution to the development of anthroposophy, and has acquired a Chestahedron to be a permanent part of Dornach's collection at the Goetheanum. In 2012 Camphill Ghent purchased a Venus Bell as a centerpiece for their new senior living community in New York, and the Ruskin Mill Educational Trust helped Freeman College cast a Chestahedron Bell for their own grounds in England. The decatria is being used as a template for a new kind of beehive form, and Frank has inspired new endeavors via a research residency program offered through the Threefold Educational Center.

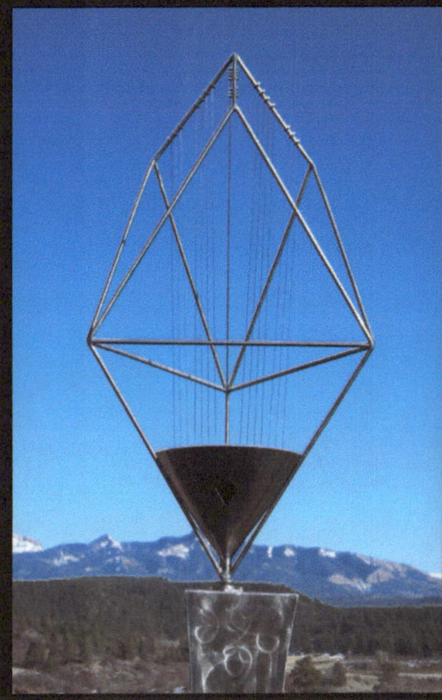

The Chesta-Tetra Wind Harp

Frank, who conducts dynamic and engaging lectures and workshops across the globe, was invited to share his discoveries both publicly and privately at the nation's premier design school, the Rhode Island School of Design, which subsequently acquired a Chestahedron to be a part of their permanent collection. His work has also inspired new artistic explorations, for example in Ross Barrable's creation of a large wind harp based on the Chestahedron. In my own work with the form, I discovered that the geometry of the Chestahedron, despite its unique shape, can be used to make a complete ring with 96 Chestahedra, a linear column, and rings of either 4 or 8 Chestahedra, as well as other fascinating and unexpected forms.

The diversity of these sources of recognition paints a picture of an unusual confluence of artistic, spiritual, and scientific connections engendered by Frank's discoveries. The work crosses established disciplinary boundaries with a formidable but essential challenge: to utilize transdisciplinary methods that speak to the human being as a whole in a multi-layered, evolving cosmos.

Matt Taylor has used the decatria to create a new shape for constructing beehives. Inside the decatria, formed out of a cross-section, is a perfect hexagon, which is the natural organizing shape for honeycomb construction.

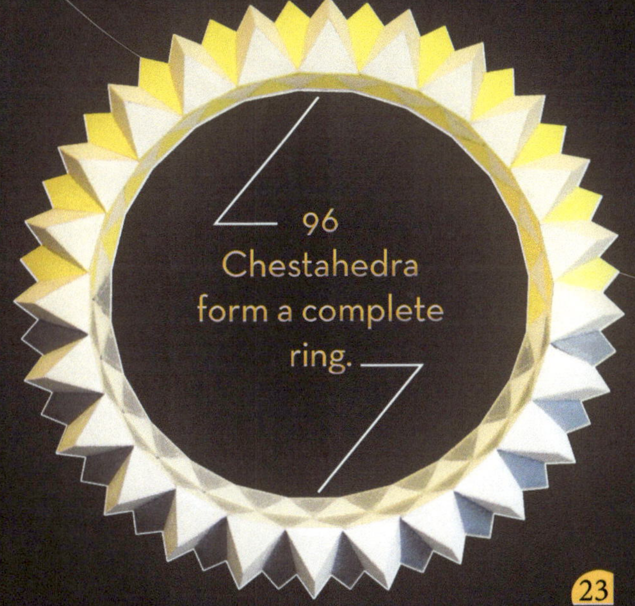

96 Chestahedra form a complete ring.

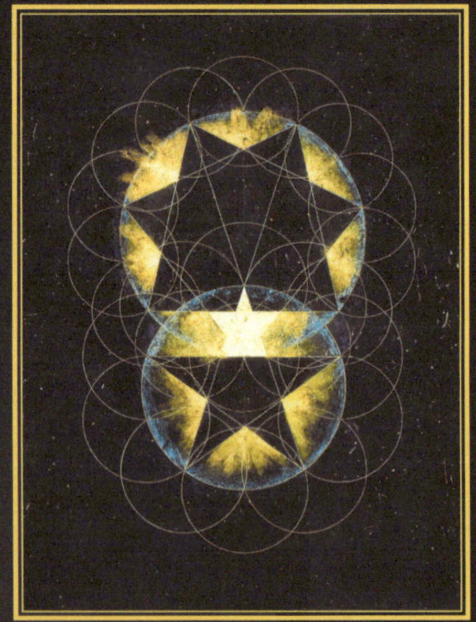

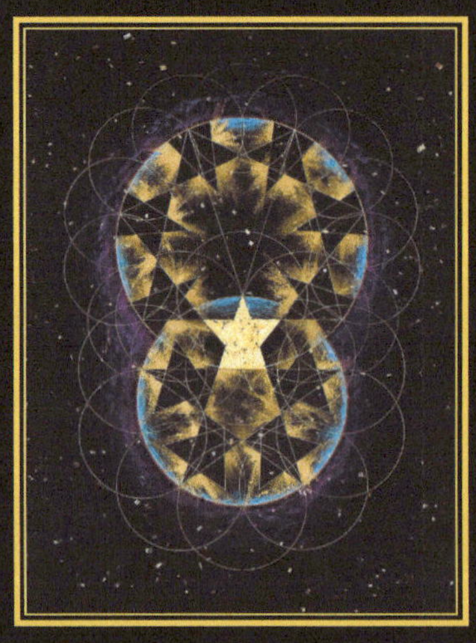

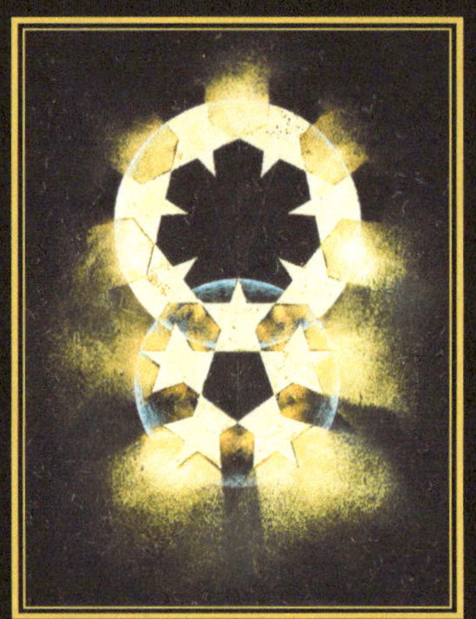

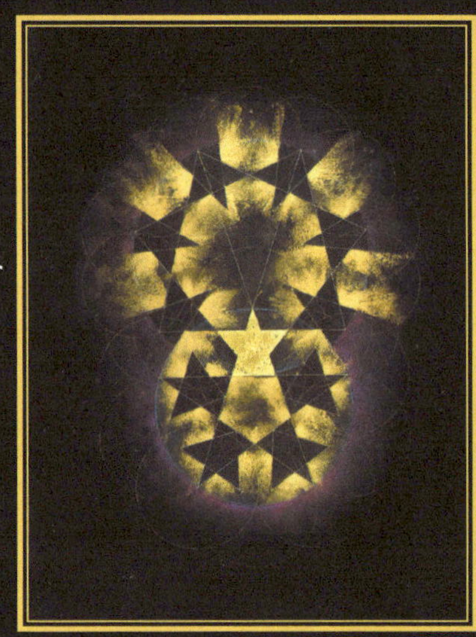

The sale of paintings, sculptures, and jewelry supports the ongoing development and application of Frank's work.

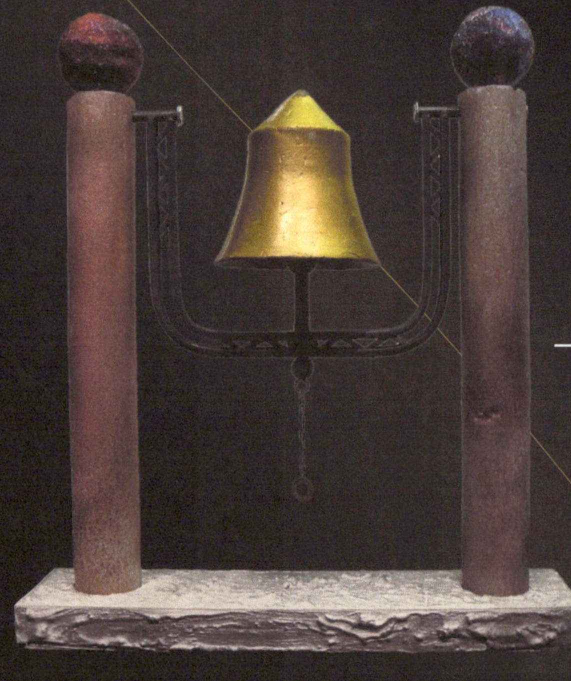

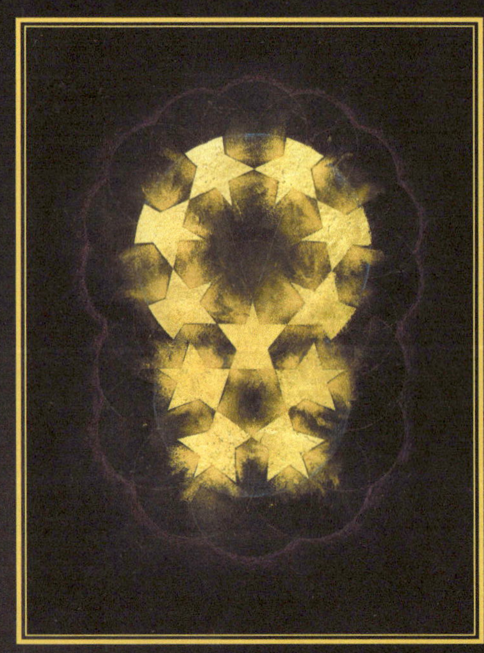

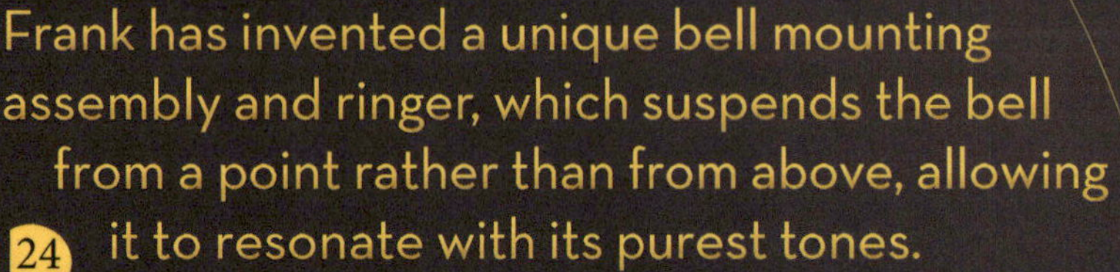

Frank has invented a unique bell mounting assembly and ringer, which suspends the bell from a point rather than from above, allowing it to resonate with its purest tones.

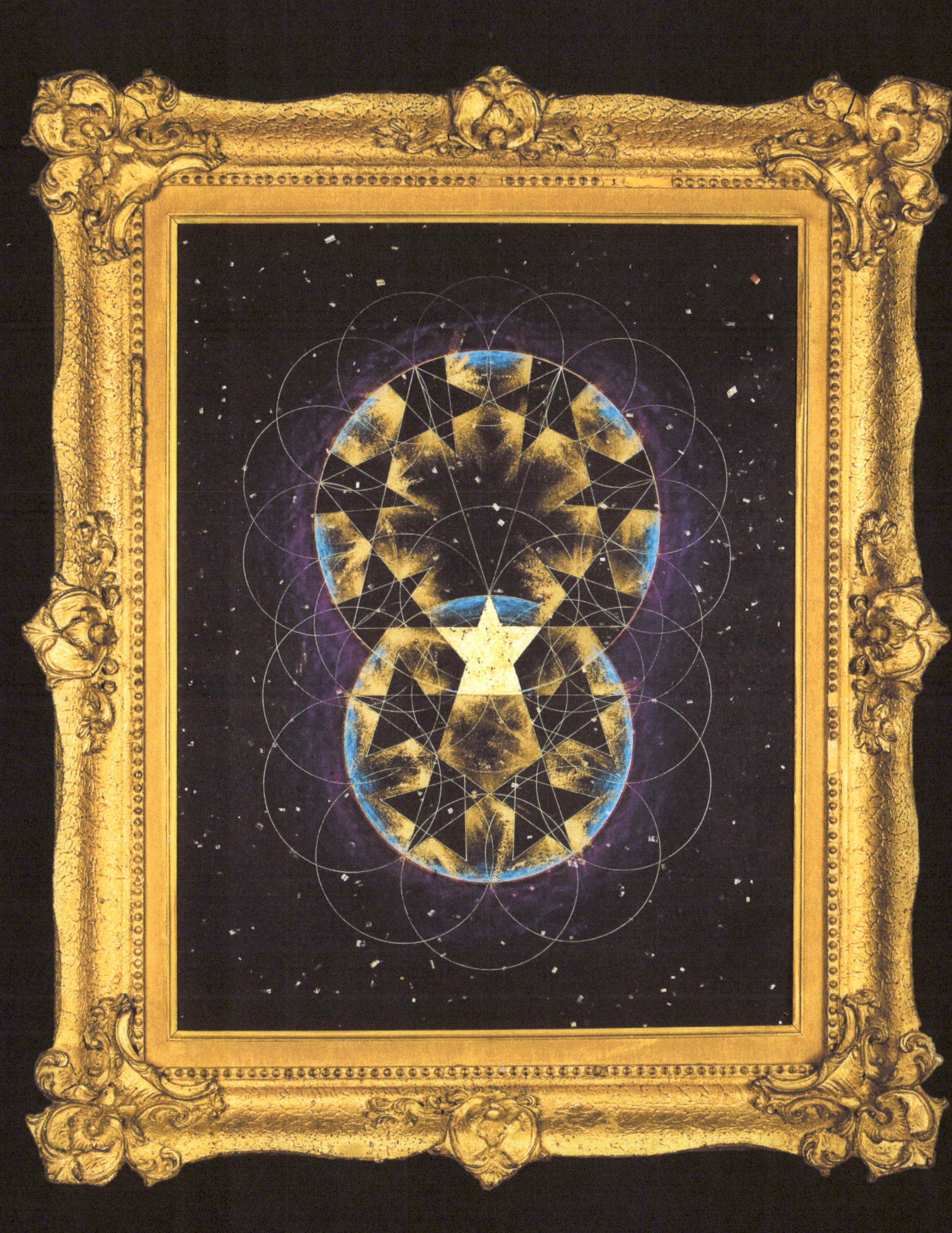

Chestahedron Studies

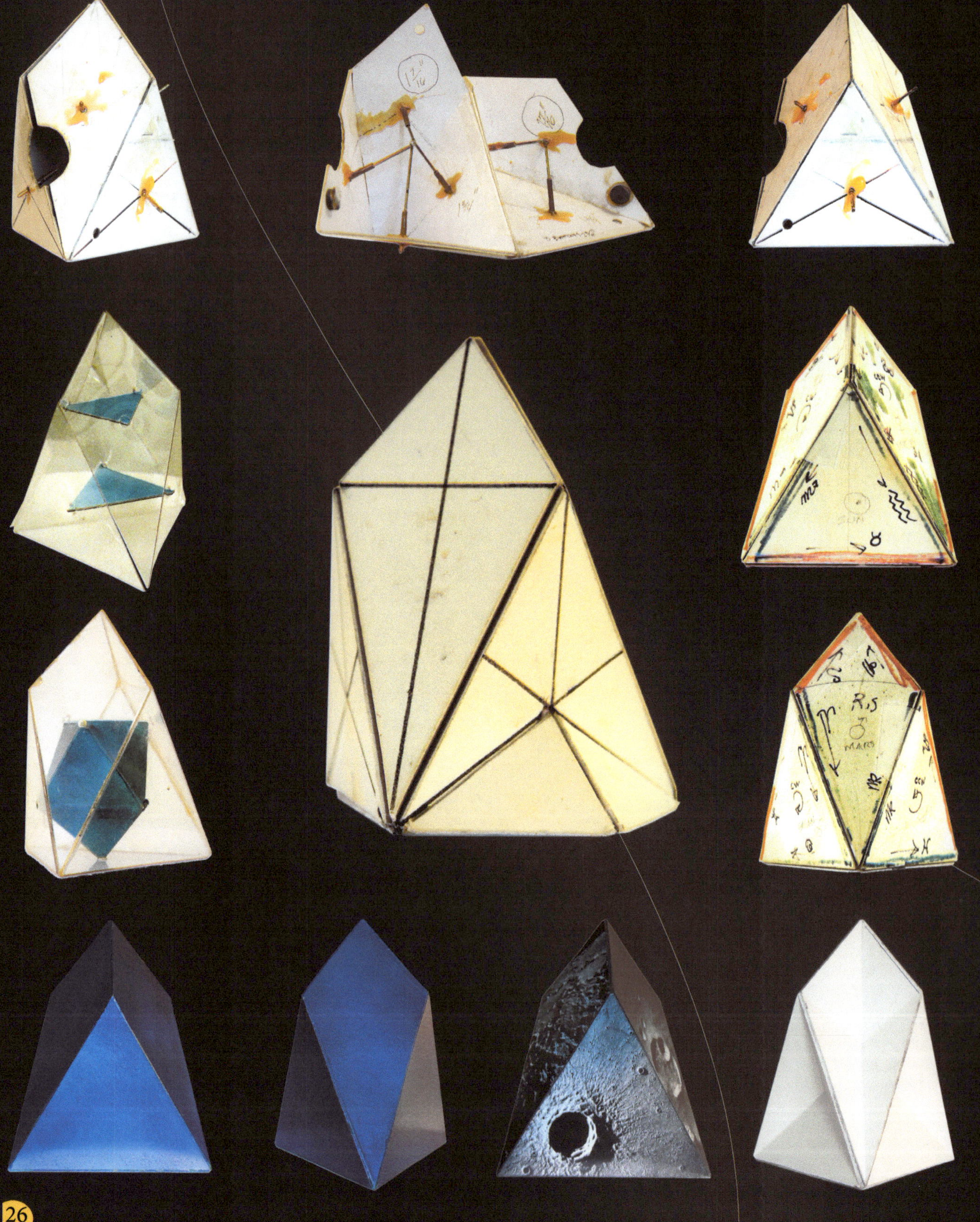

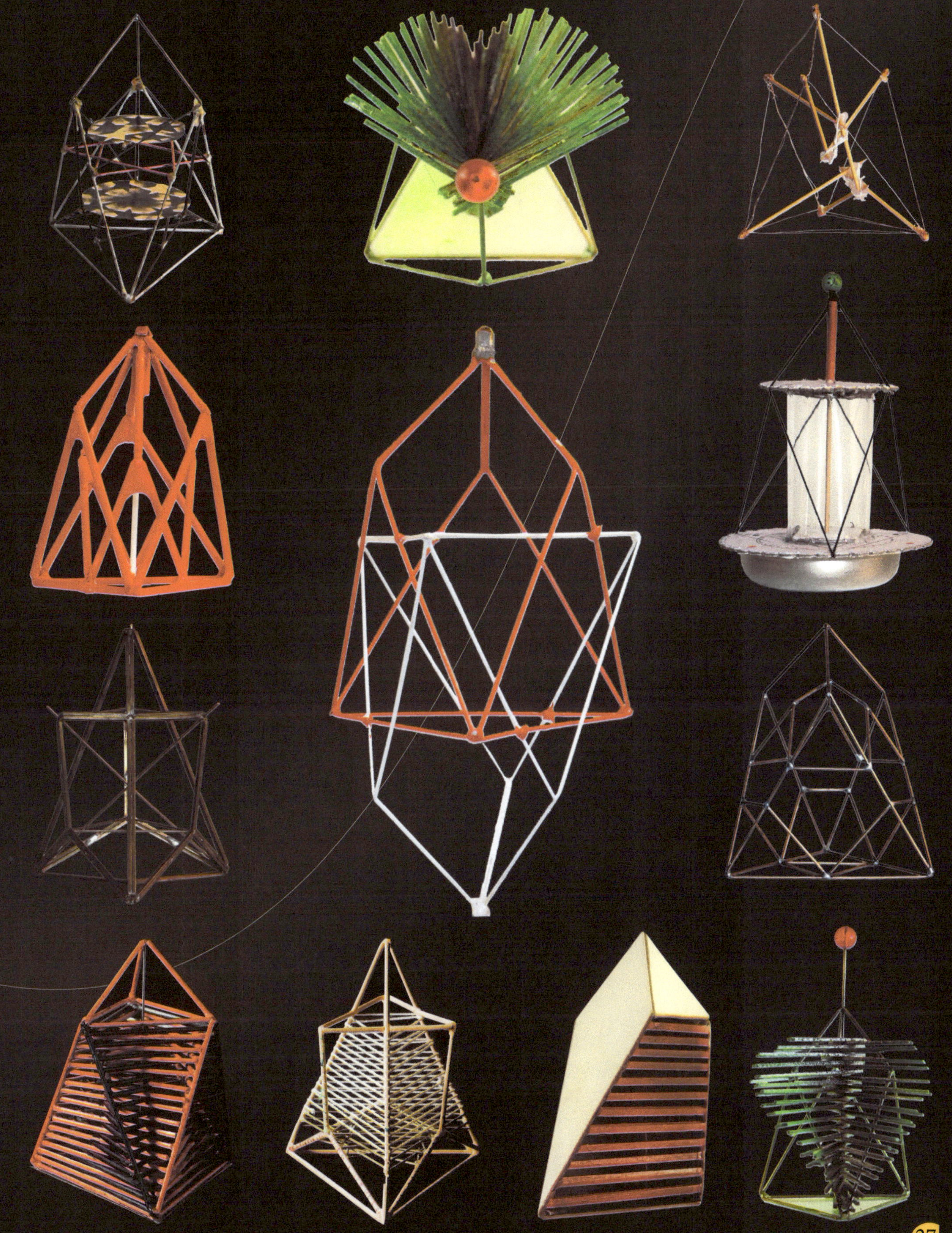

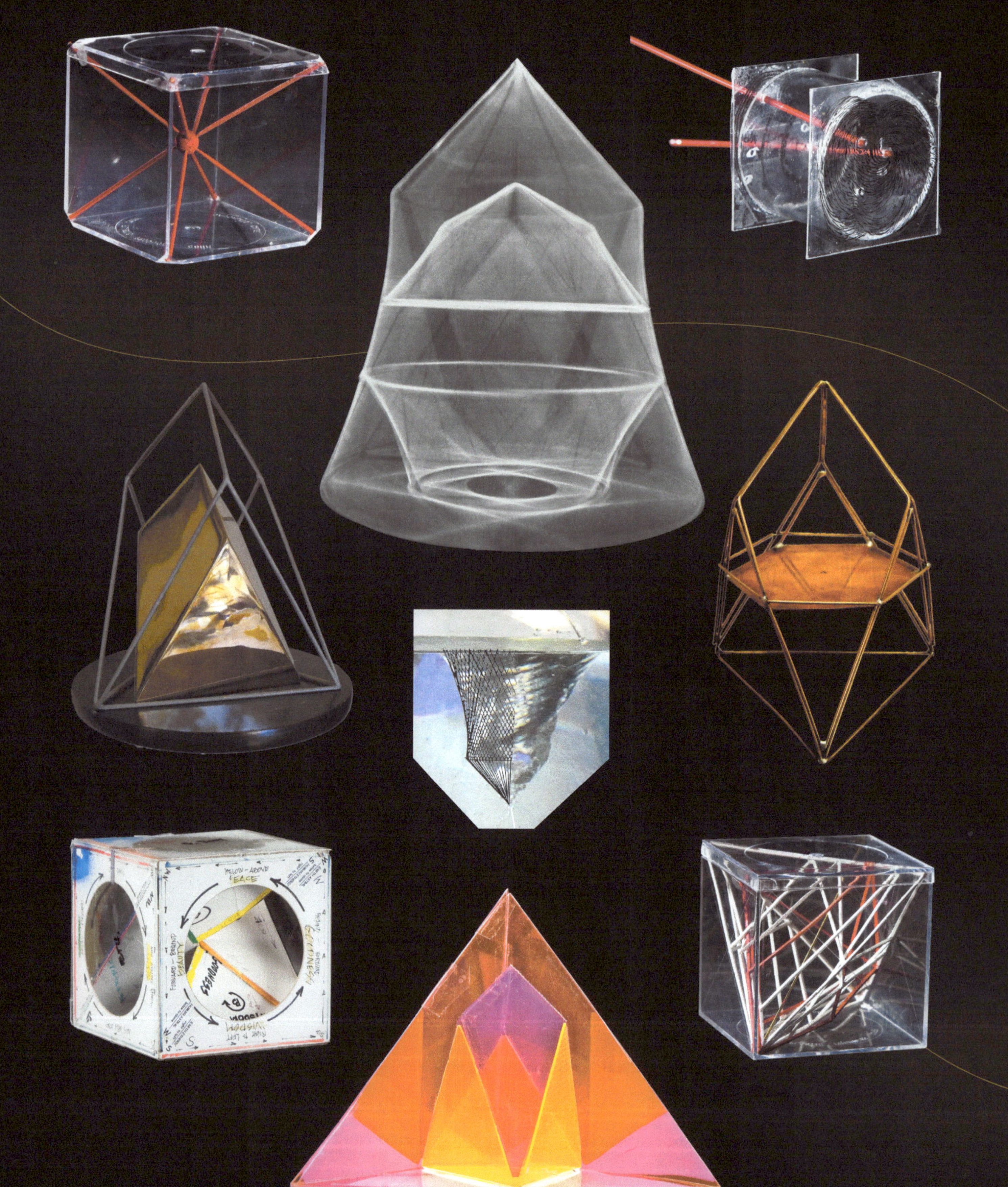
BELOW – ABOVE
PEACE
BEHIND – FORWARD
GOODNESS
FORWARD – BEHIND
BEAUTY
RIGHT TO LEFT
WISDOM

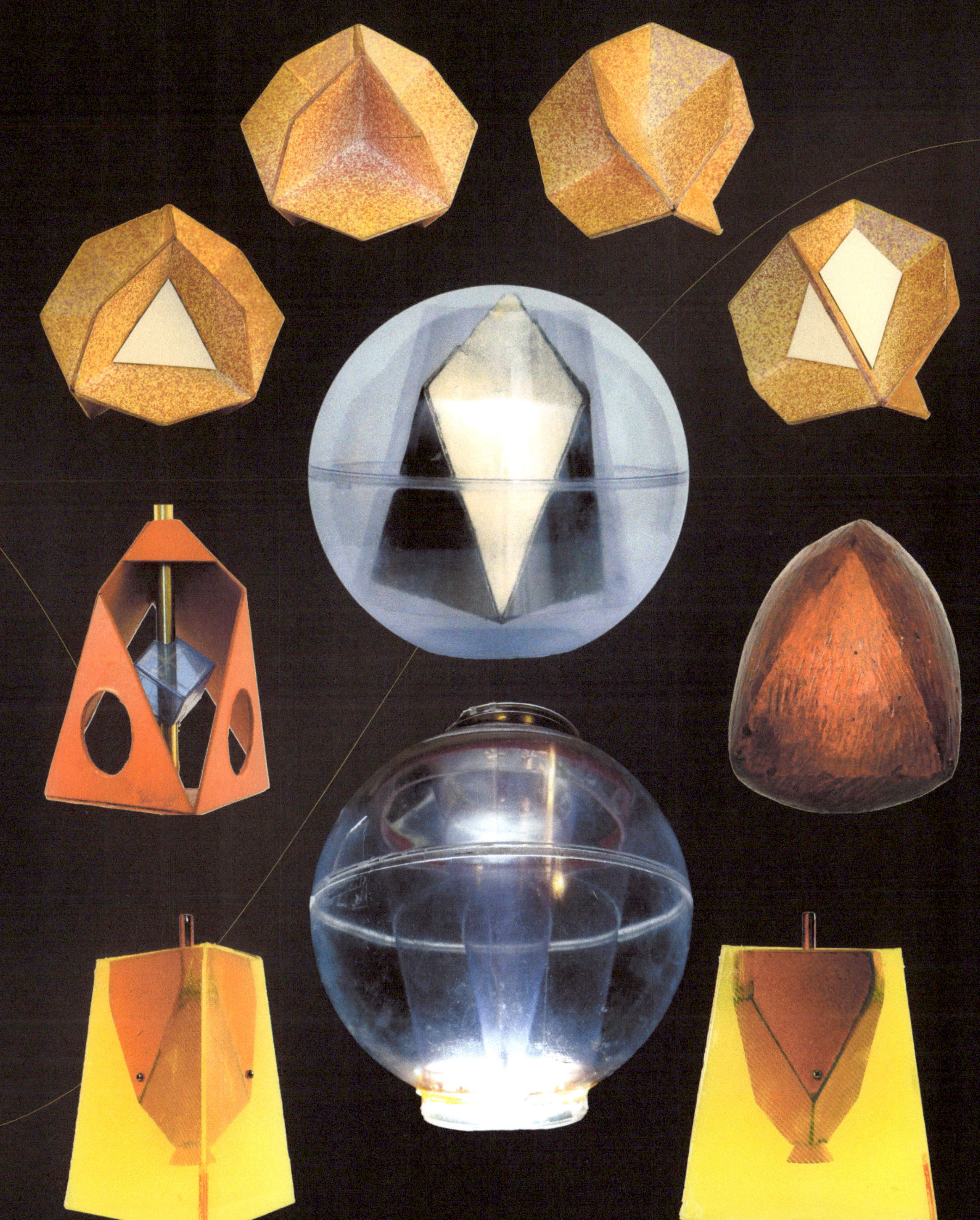

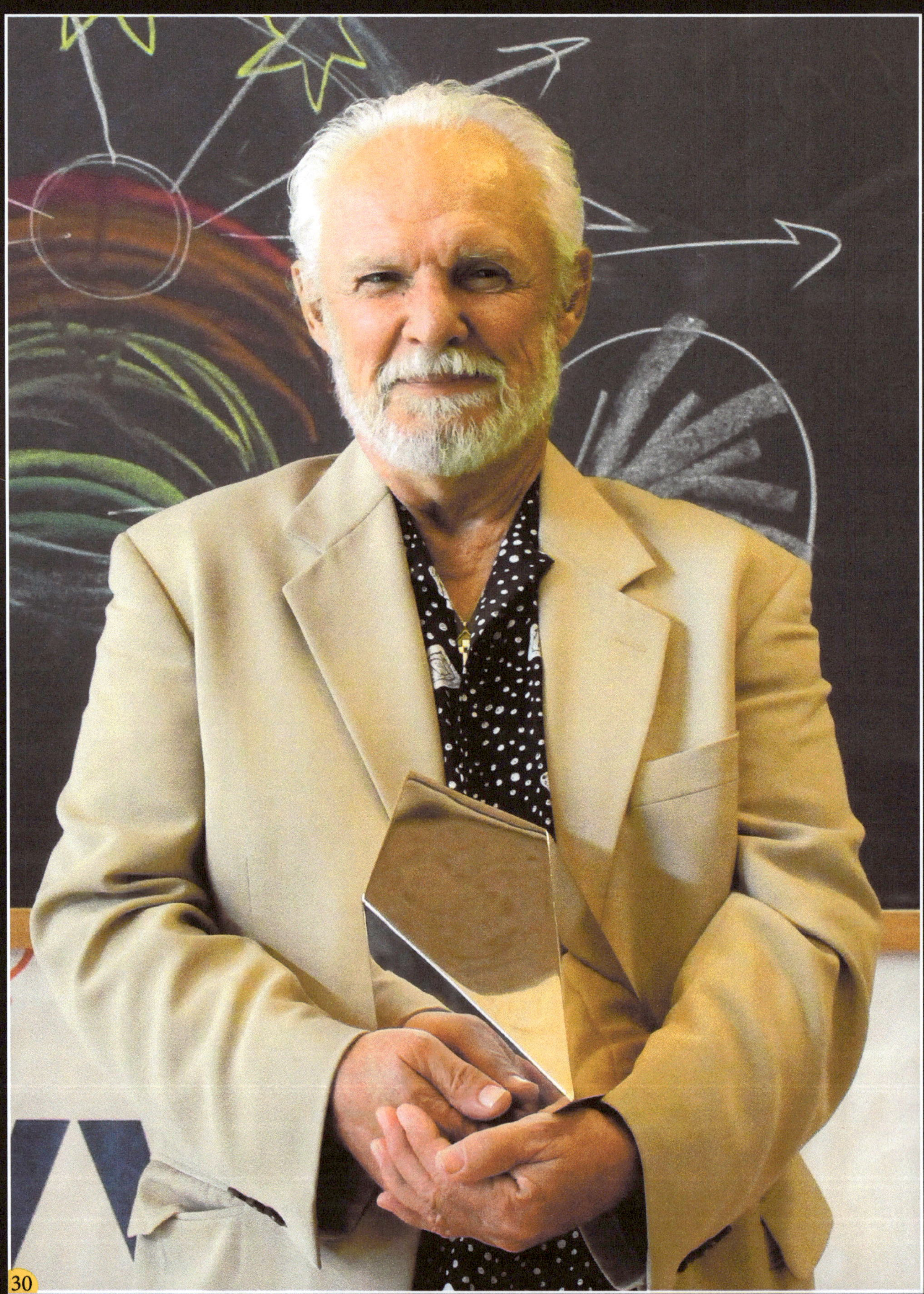

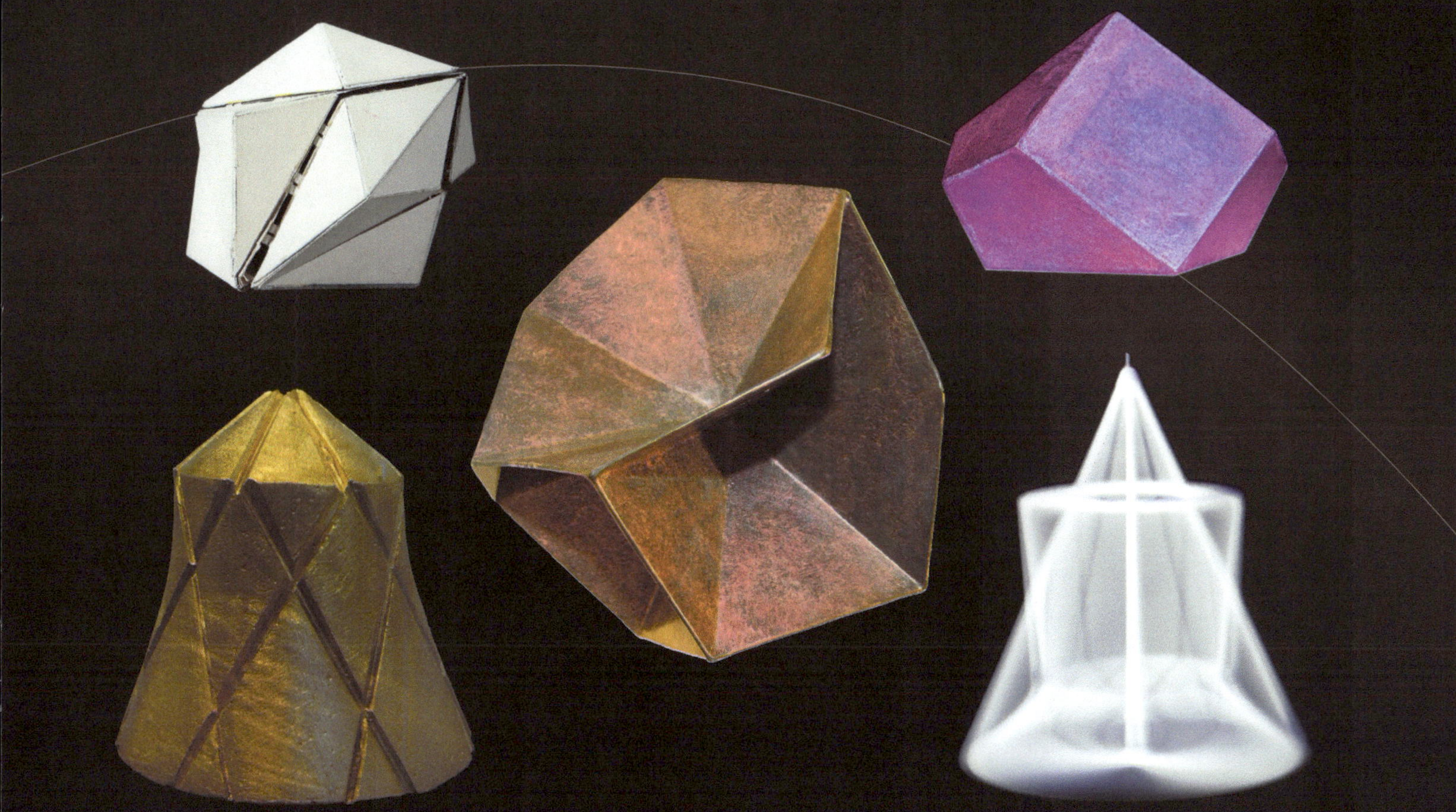

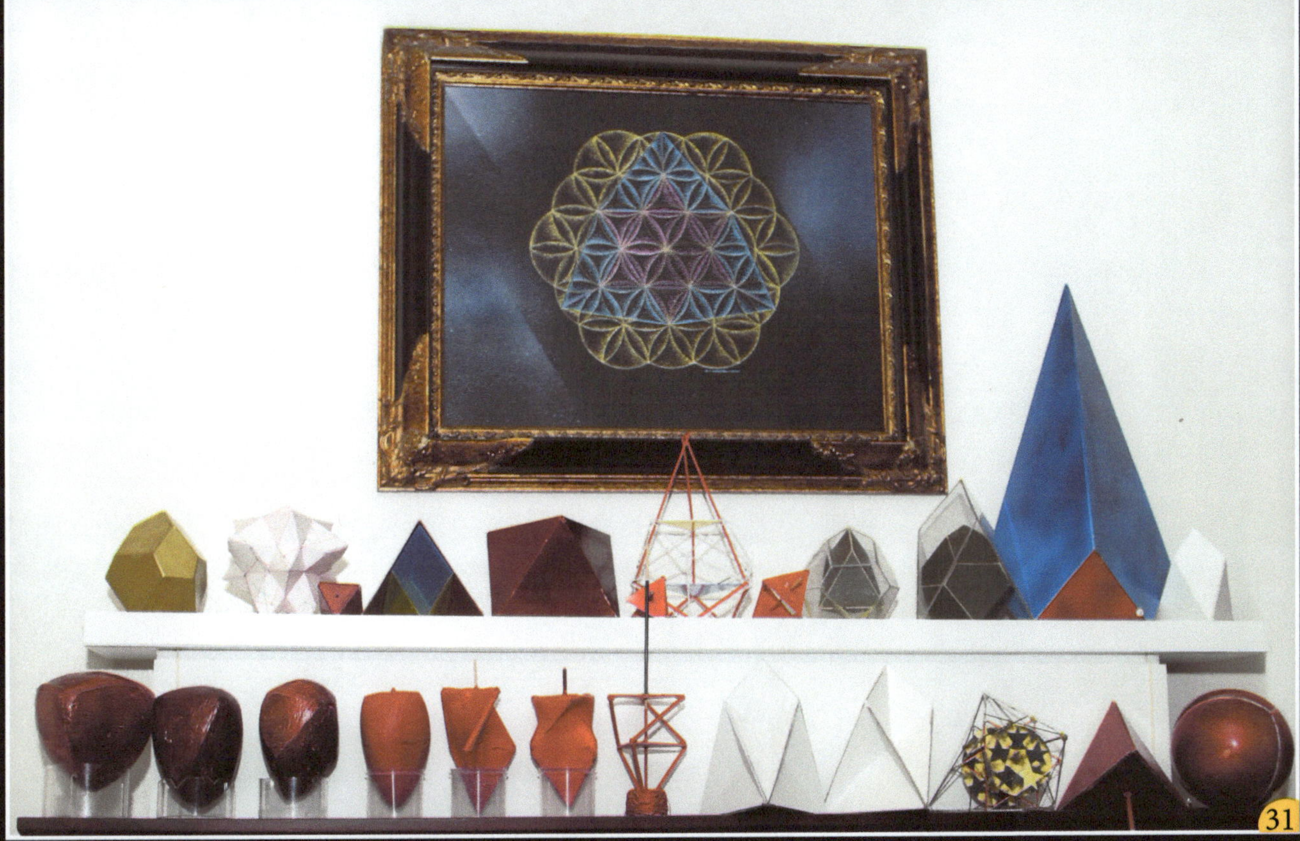

Frank Chester has also invented a device, designed upon the geometry of the Chestahedron, called the Chesta Vortex Organizer (CVO). The CVO is meant to be utilized in biodynamic preparations, stirring, mixing, aeration and bio-augmentation. The device has two moving forms turning in opposite directions, 90 degrees from each other.

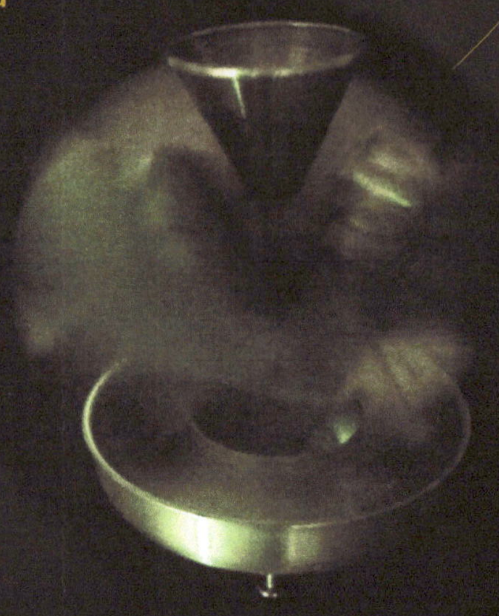

Frank has been given the opportunity to test the new device at Environment Ventures Marketing, Inc. in the Philippines. He has been invited to a professional laboratory environment which includes scientists, engineers, fabricators, and machines to produce and test many of his ideas.

The dualization of the Chestahedron into the Decatria, and the Decatria into the Chestahedron.

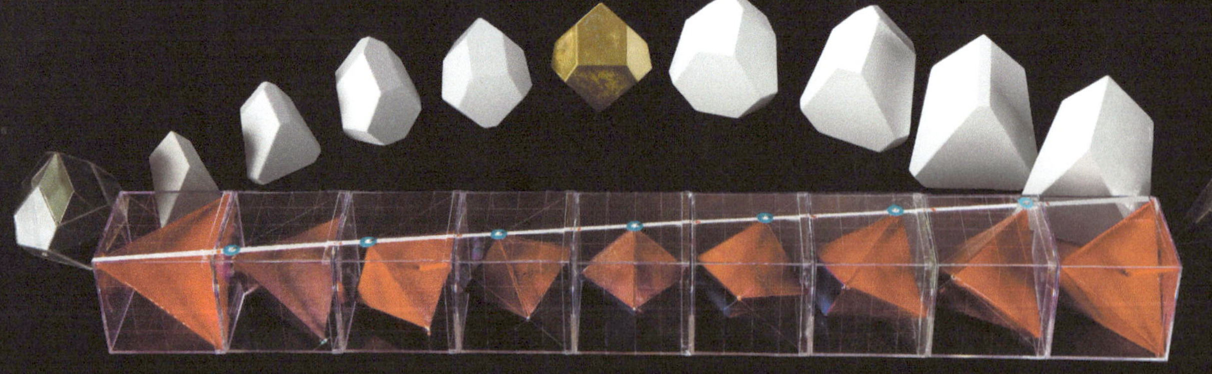

One way that the Chestahedron can be formed is by rotating a tetrahedron within a cube.

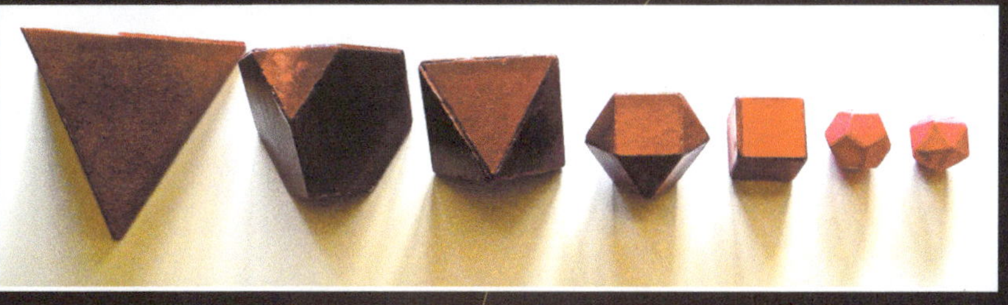

The phenomenological sequencing of the Platonic forms, in scale to each other.

The morphological generation of the seven-sided form as an archetype by opening the tetrahedron, with the Chestahedron as the balance point.

The decatria in the Chestahedron.

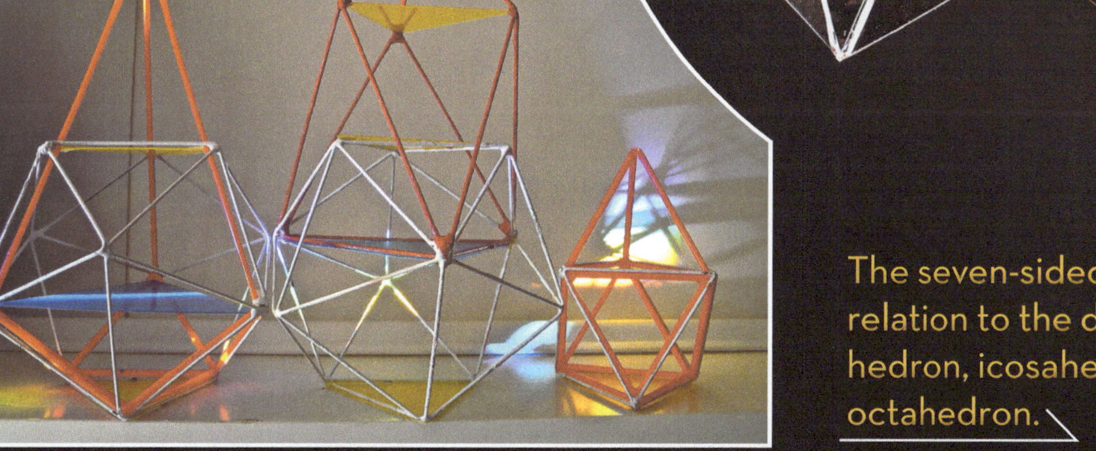

The seven-sided form in relation to the dodecahedron, icosahedron, and octahedron.

A selection of Frank's geometric sculptures based on the Chestahedron.

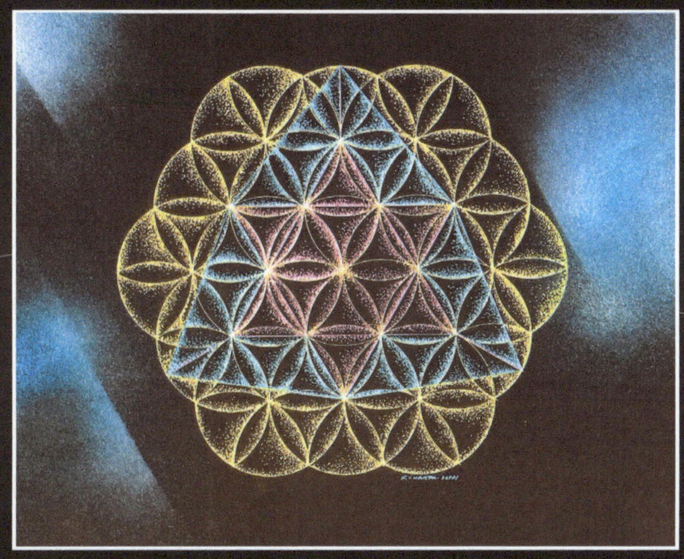

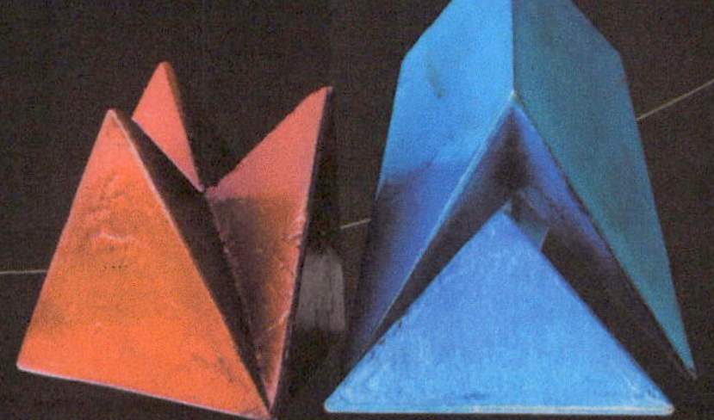

Dissecting the Chestahedron

The Chestahedron's geometry integrates with the Flower of Life

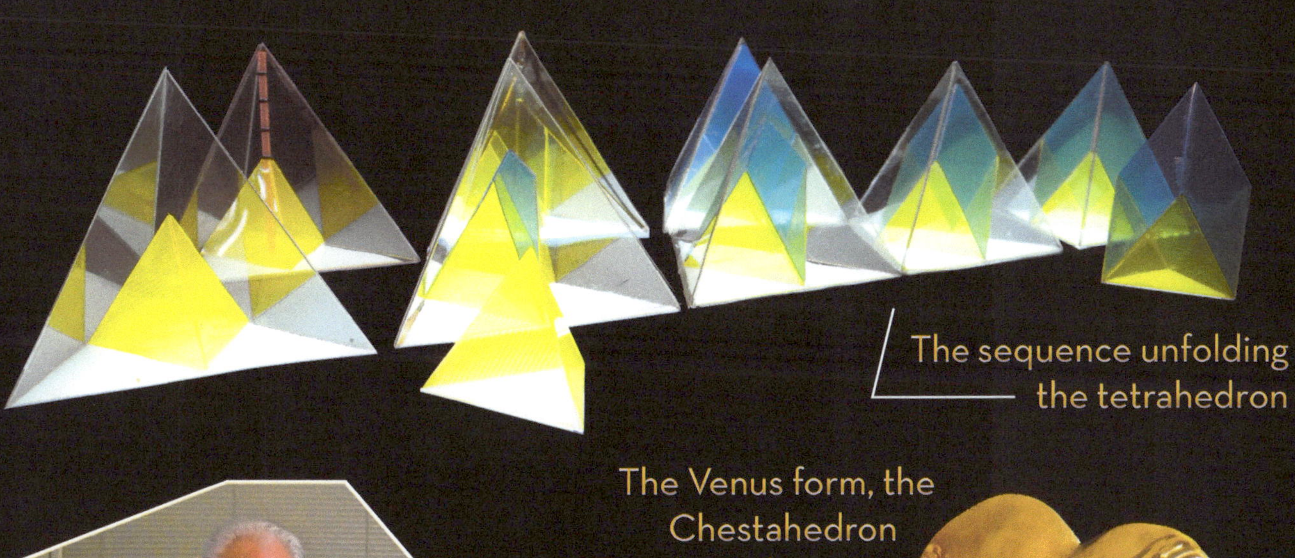

The sequence unfolding the tetrahedron

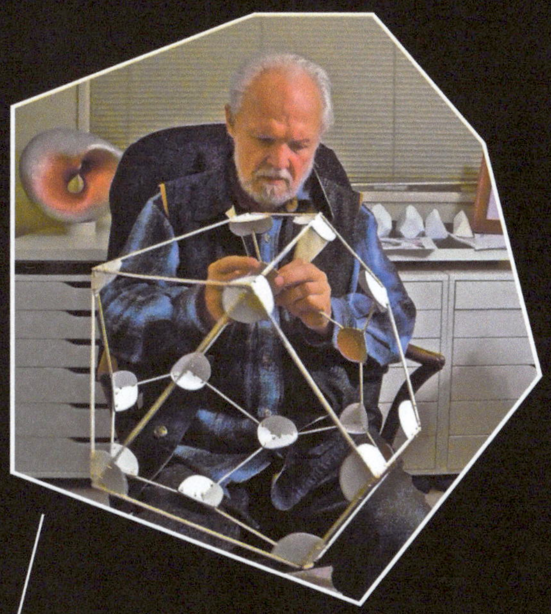

The decatria is a potential alternative to geodesic dome construction

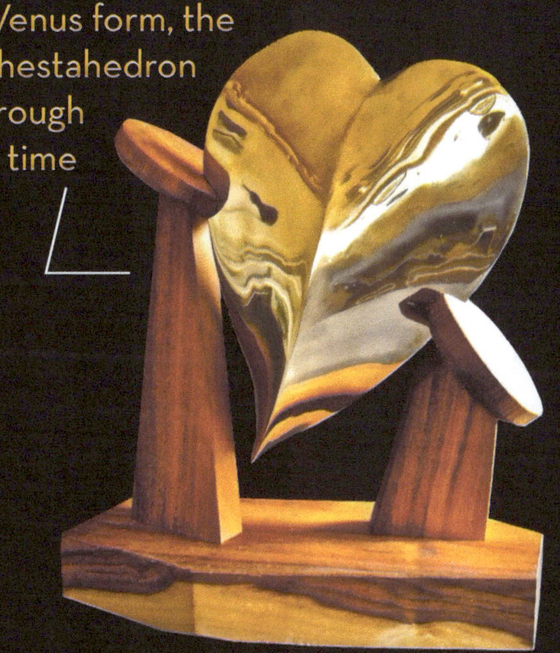

The Venus form, the Chestahedron through time

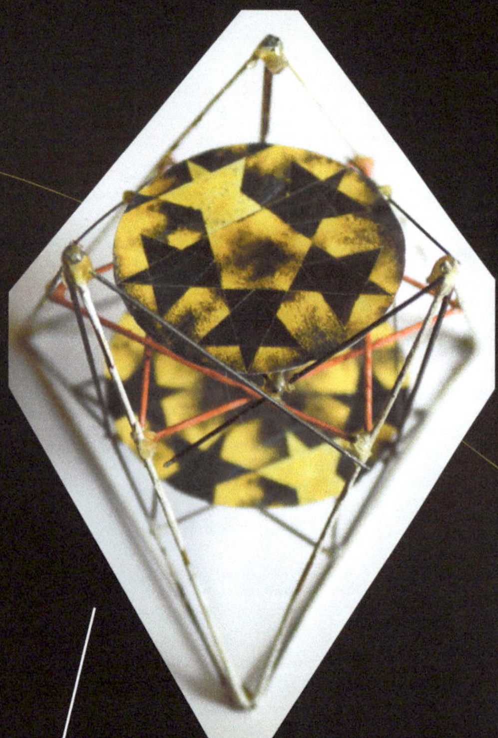

A perfect hexagon slices (in red) through the Chestahedron

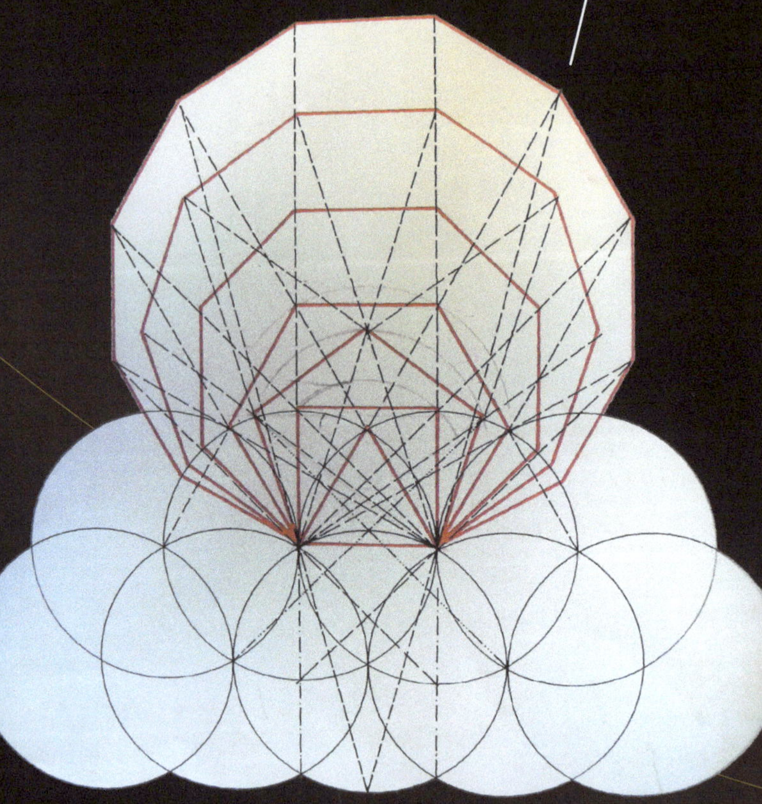

Relationships between successive polygons

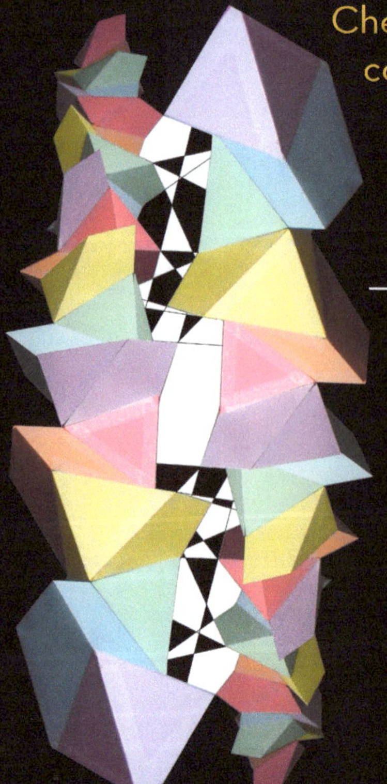

Chestahedra can be connected to form a linear column

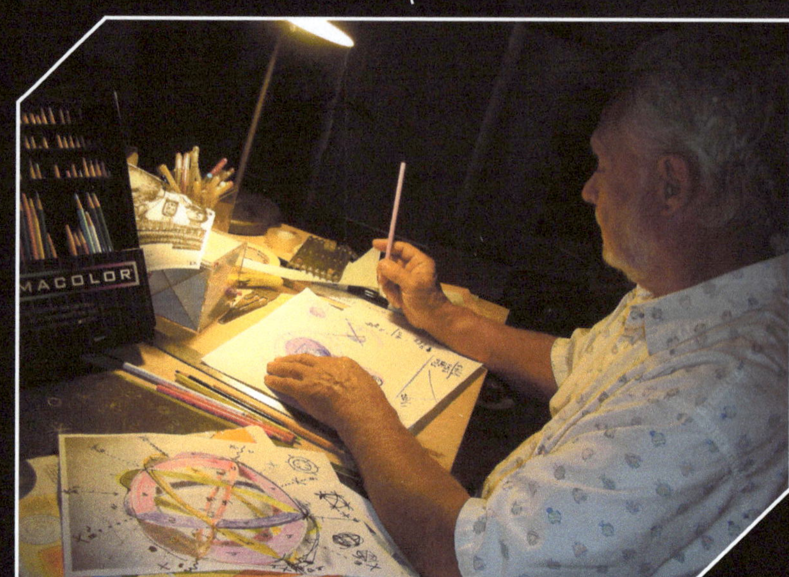

Frank hard at work on new geometry

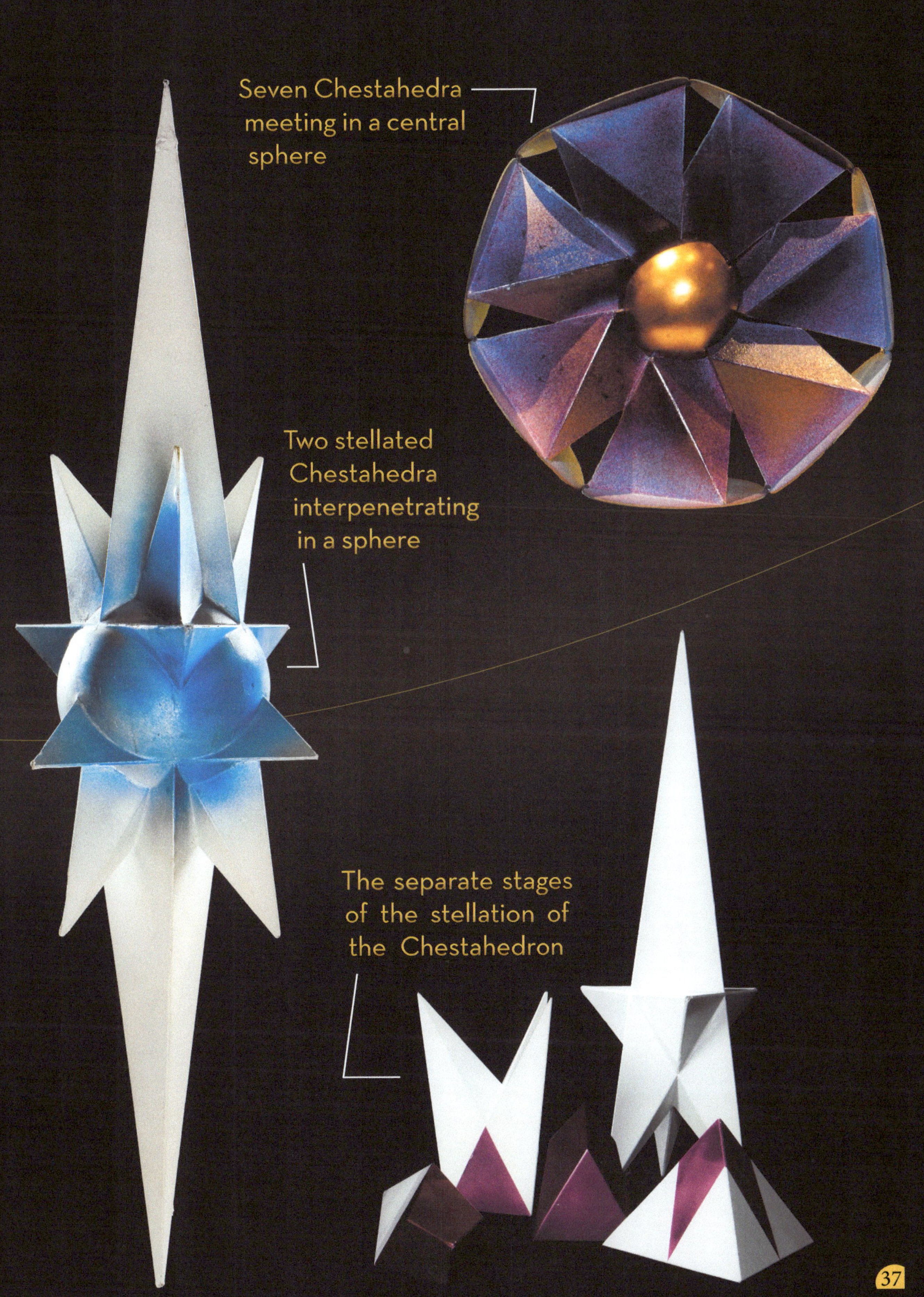

Seven Chestahedra meeting in a central sphere

Two stellated Chestahedra interpenetrating in a sphere

The separate stages of the stellation of the Chestahedron

Squaring the Circle Construction Method

Frank has discovered a new way to square the circle so that the perimeter of the square and the circumference of the circle are almost exactly equal. His method uses only a straight-edge and compass, and is the first to work from the inside out using the Vesica Piscis. The protocol is as follows (refer to the image on the following page):

1. Draw a circle (#1).
2. Open your compass to the diameter of circle #1, and draw a second circle (#2) (that has a radius exactly twice the original circle), whose center lies on the perimeter of circle #1.
3. Draw a light line (m) connecting the center of circles #1 and #2, projecting it across most of the page.
4. Use where this line m meets the opposite side of circle #1 to be the center of a new circle, with the same radius as circle #2, this is circle #3.
5. Circles #2 and #3 form the Vesica Piscis, and intersect each other in two new points. Connect these points with a line (n), which is orthogonal to line m.
6. Where line n intersects circle #1 it creates two new points. Use these as the centers of two new circles with the same radius as Circles #2 and #3; let point Q be the center of circle #4 and point R be the center of circle #5. These new circles, #4 and #5, form another Vescia Piscis at 90 degrees to the first. The two Vescia Piscis cradle circle #1 where they overlap.
7. Circles #4 and #5 intersect in two points that lie on line m. Choose one of these points (P), and draw a line (s) orthogonal (at 90 degrees) to line m (and thus parallel to line n) at this point. From point P, extend line s on either side of line m so that its total length is just larger than the diameter of circle #1.
8. Note the center points of circles #4 and #5, Q and R. From each of these center points draw a line perpendicular to line n (parallel to line m), extending them to meet line s in two new points (S and T). Let S be the point made by extending from point Q and T be the point made from extending point R.
9. Open your compass to the length made between points S and R, and draw a circle (#6) of that radius with its center on point R.
10. Draw another circle (#7) of the same radius, with its center on point Q.
11. Circles #6 and #7 intersect line n in two new points (U and V) towards the outsides of the construction. Open your compass to the length made between the center of circle #1 and one of these new points, and draw a new circle (#8) with the same center as circle #1.
12. Circle #8 meets lines m and n in four points. Construct a total four lines (one on each point), two parallel to n and two parallel to m, such that the lines form a square in which circle #8 is circumscribed.
13. Open your compass to the original radius of circle #1. With the point of your compass on point U, mark a new point (Y) on line n towards the outside of the construction.
14. Draw a new circle (#9) with a center on point Y, with a radius equal to circle #1. Circle #9 is tangent to circle #8.
15. Open your compass to the distance between the centers of circles #1 and #9, and draw a final circle (#10) with the same center as circle #1. This circle has a perimeter almost exactly equal to the square constructed in step 12.

Perimeter of the square = Circumference of circle #10

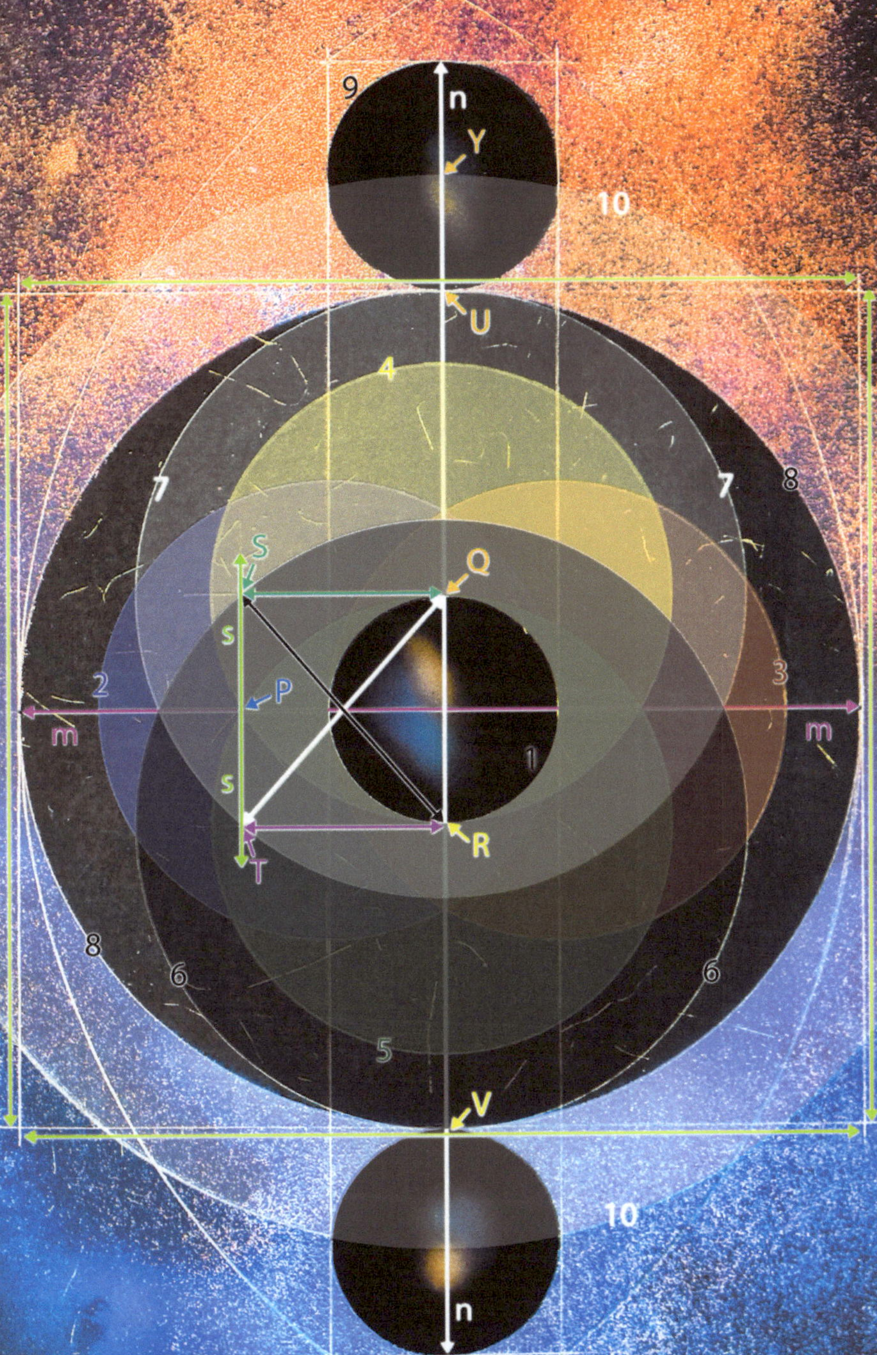

Squaring the Circle using the Vesica Piscis

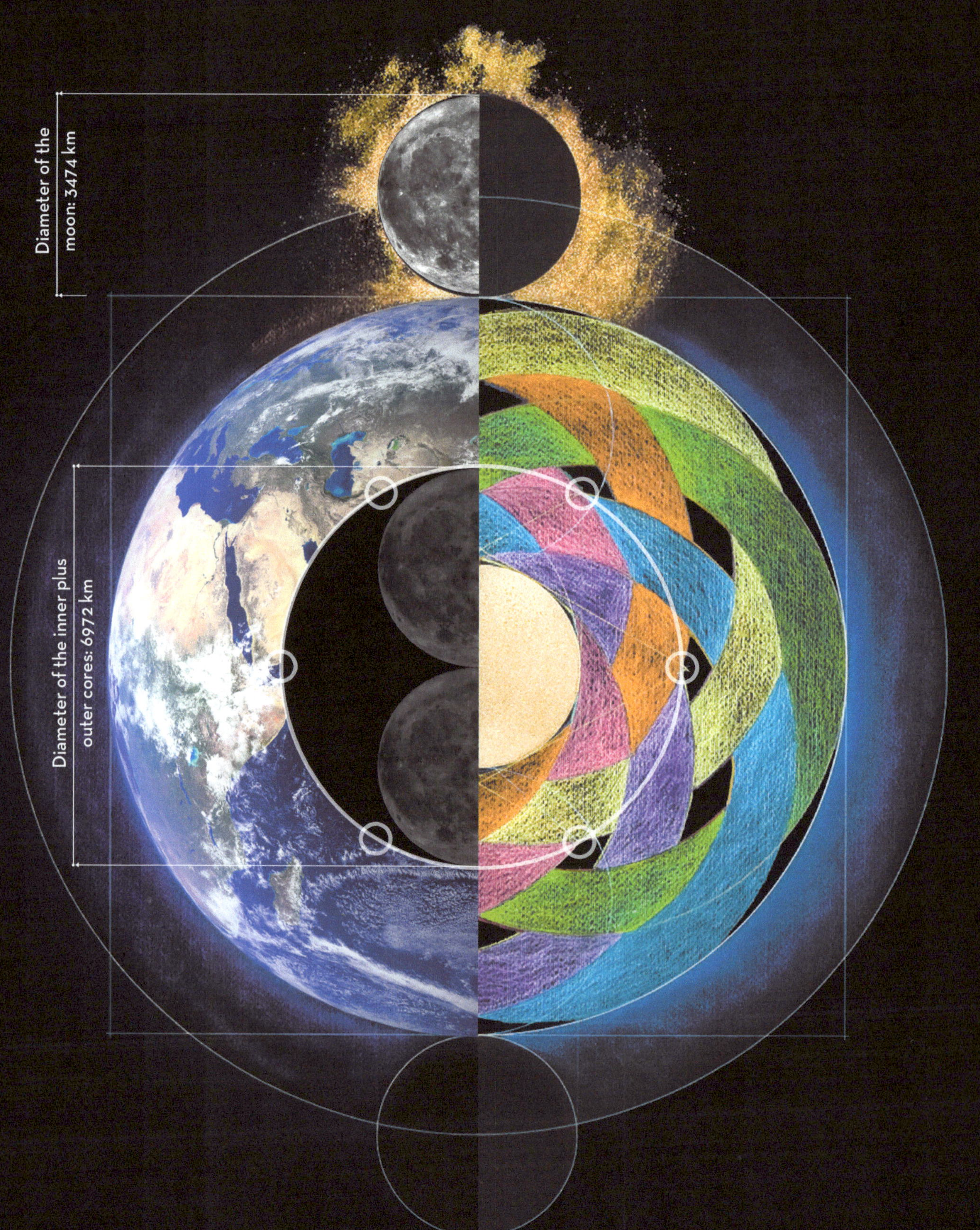

HEPTAHEDRON FROM THE VESICA PISCIS

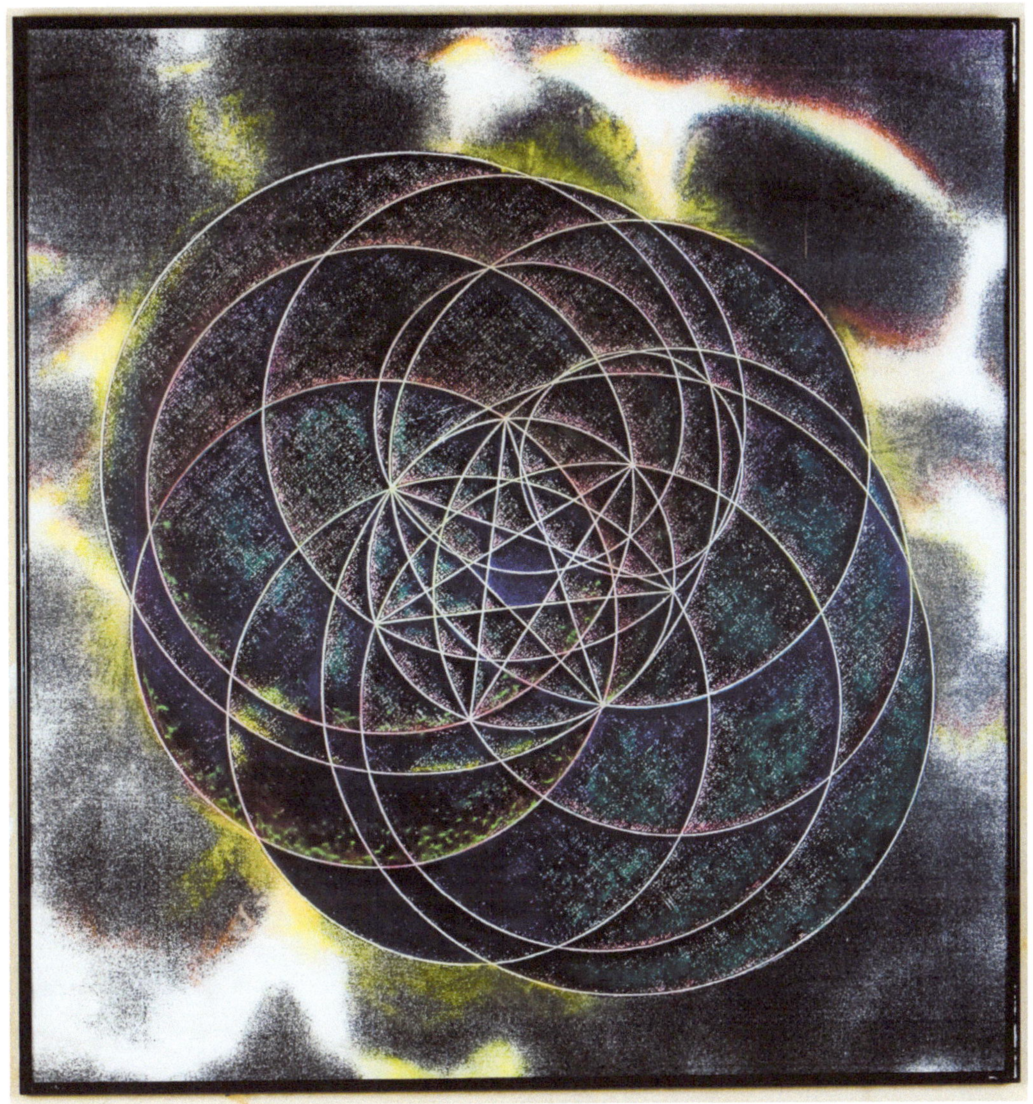

Frank has also discovered a new straightedge and compass procedure to create a heptagon, a seven-sided polygon, using the Vesica Piscis. This is the first known construction that places the heptagon *inside the center* of the Vesica Piscis. As in the case of squaring the circle, a mathematically precise straightedge and compass construction is impossible, but this simple method comes very close.

Constructing a Heptagon

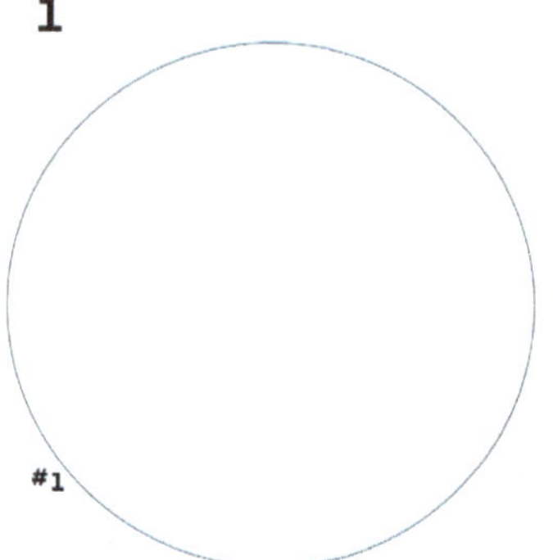

1. Draw a circle (#1).

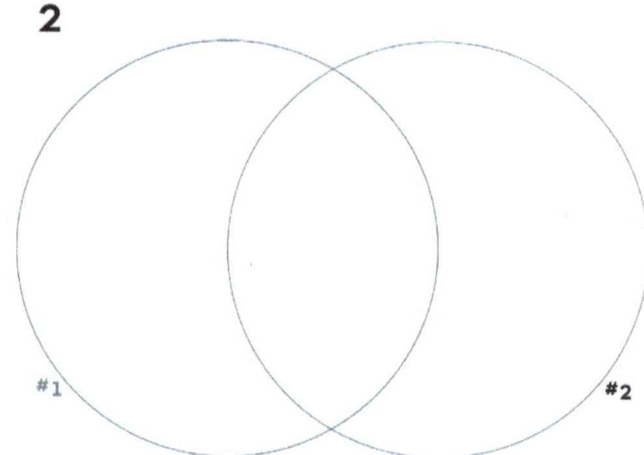

2: Draw a second circle (#2) with the same radius as circle #1, and with a center that lies on the perimeter of circle #1.

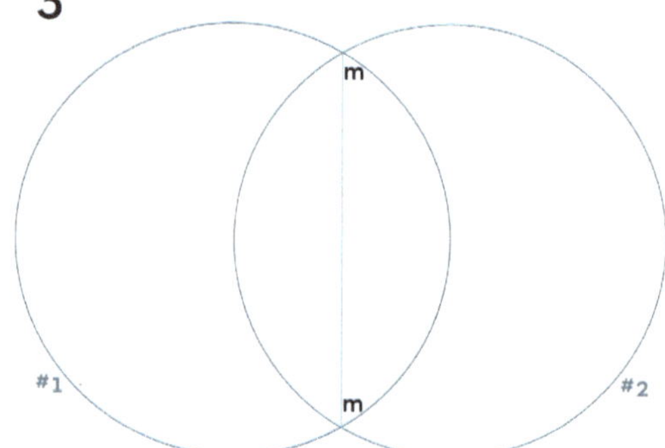

3: Draw line m connecting the two crossing points of circles #1 and #2.

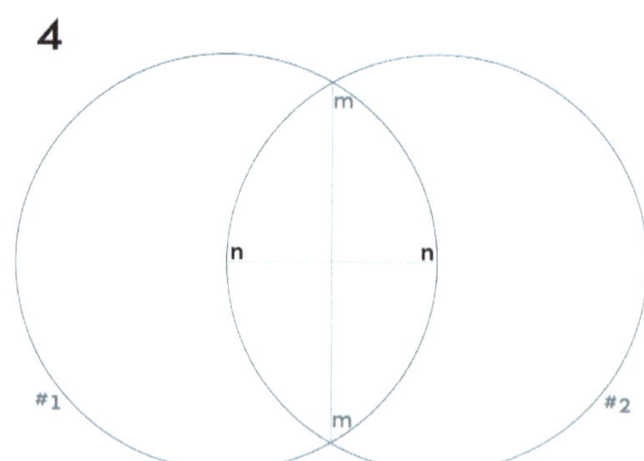

4: Connect the centers of circles #1 and #2 with line n, which is perpendicular to line m.

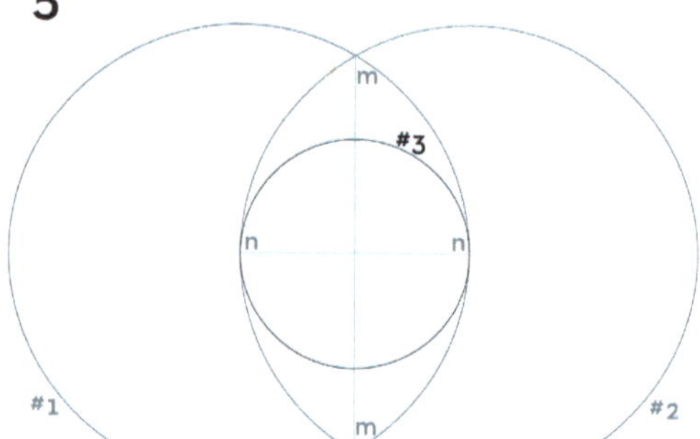

5: Place the tip of your compass on the point where lines n and m meet, and open it to the center of circle #1, and draw circle #3, which is exactly half the size of circles #1 and #2.

6

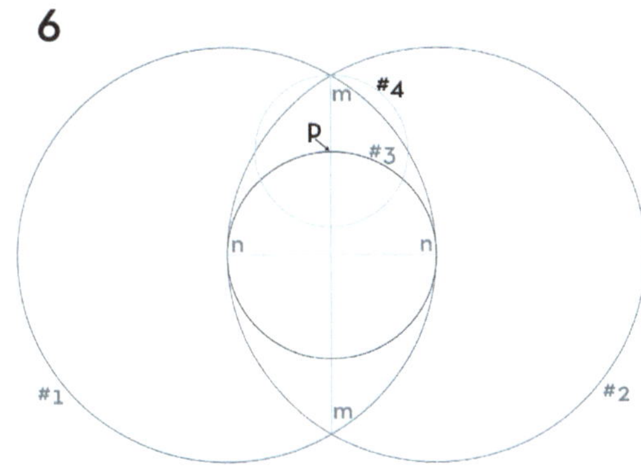

6: Circle #3 crosses line m in two places. Choose one crossing, point P, and place the tip of your compass there. This is one point of the heptagon. Open the compass to the nearby point where circles #1 and #2 meet, also on line m, and draw circle #4.

7

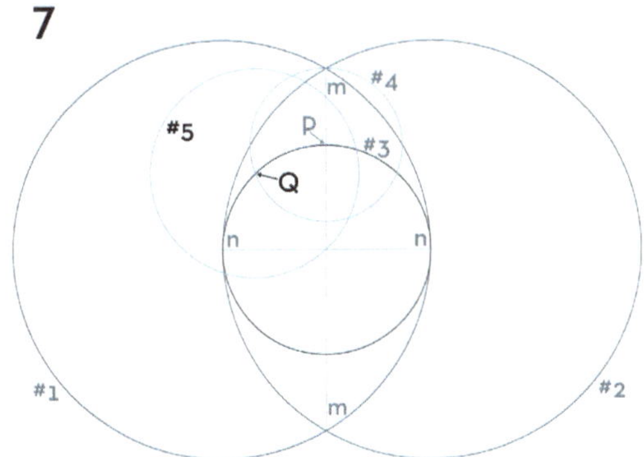

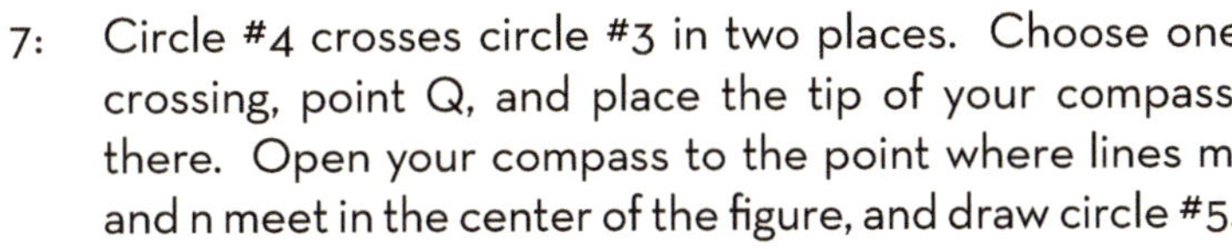
7: Circle #4 crosses circle #3 in two places. Choose one crossing, point Q, and place the tip of your compass there. Open your compass to the point where lines m and n meet in the center of the figure, and draw circle #5.

8

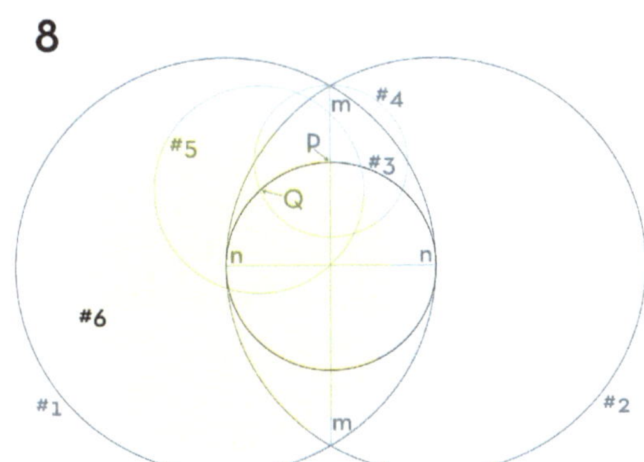

8: Circle #5 crosses circle #3 in two places, once near line m and once near line n. Place the tip of your compass on the place where circle #5 crosses circle #3 near line n. This is another point on the heptagon. Open the compass to point P and draw circle #6.

9

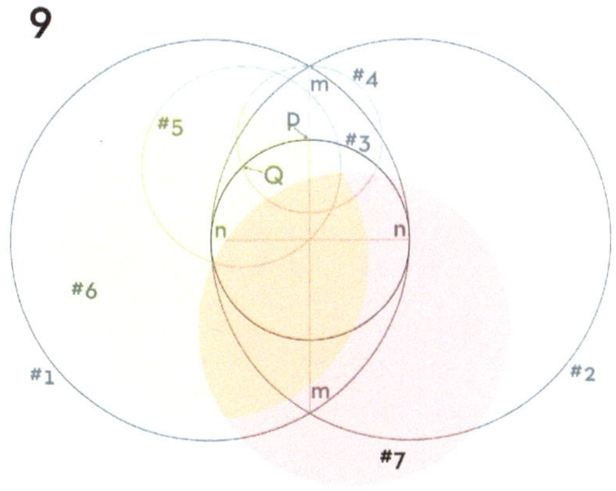

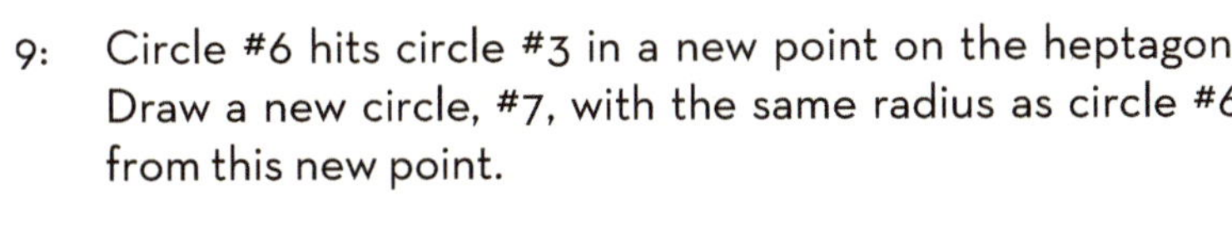
9: Circle #6 hits circle #3 in a new point on the heptagon. Draw a new circle, #7, with the same radius as circle #6 from this new point.

10

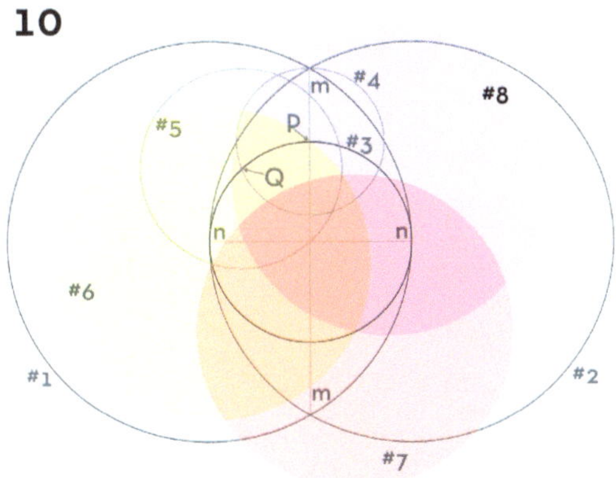

10: Circle #7 hits circle #3 in a new point on the heptagon. Draw a new circle, #8, continuing with the same radius.

11

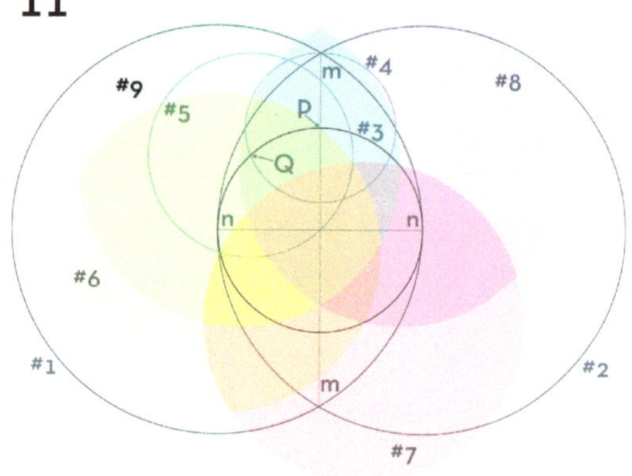

11: Circle #8 hits circle #3 in a new point on the heptagon. Draw a new circle, #9, continuing with the same radius.

12

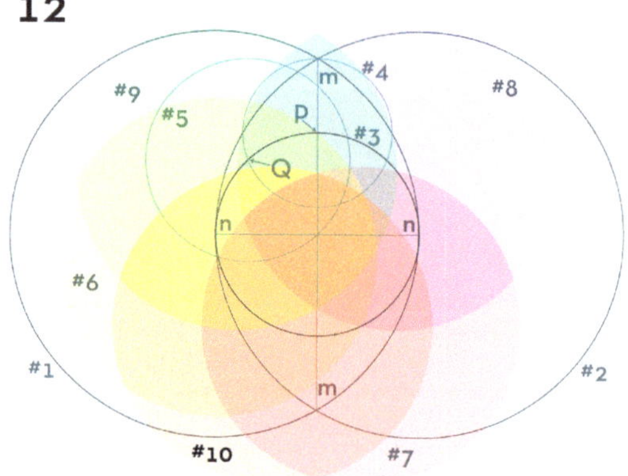

12: Circle #9 hits circle #3 in a new point on the heptagon. Draw a new circle, #10, continuing with the same radius.

13

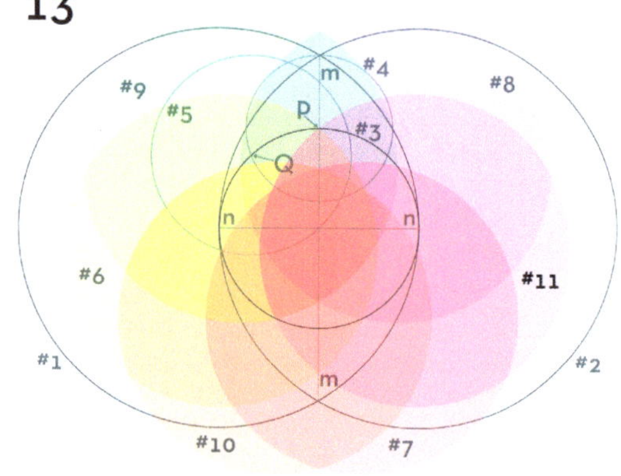

13: Circle #10 hits circle #3 in a new point on the heptagon. Draw a new circle, #11, continuing with the same radius.

14

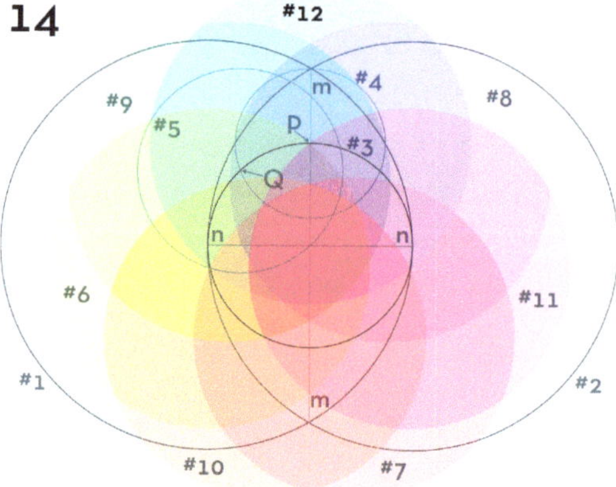

14: Circle #11 hits circle #3 in a new point on the heptagon. Draw a new circle, #12, continuing with the same radius.

15

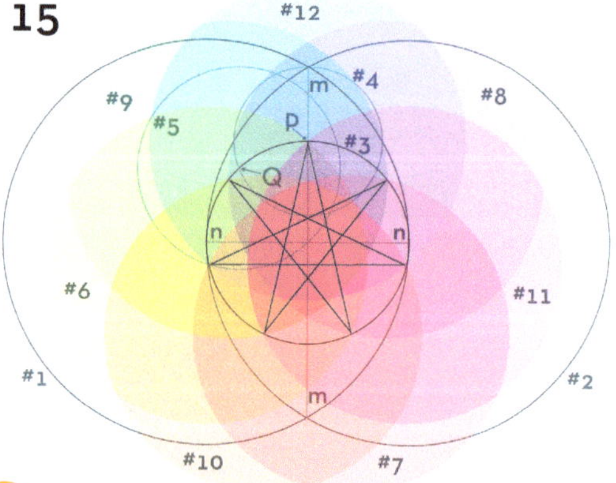

15: Connect the seven points on circle #3 into the heptagon.

Nine Vesica Pisces

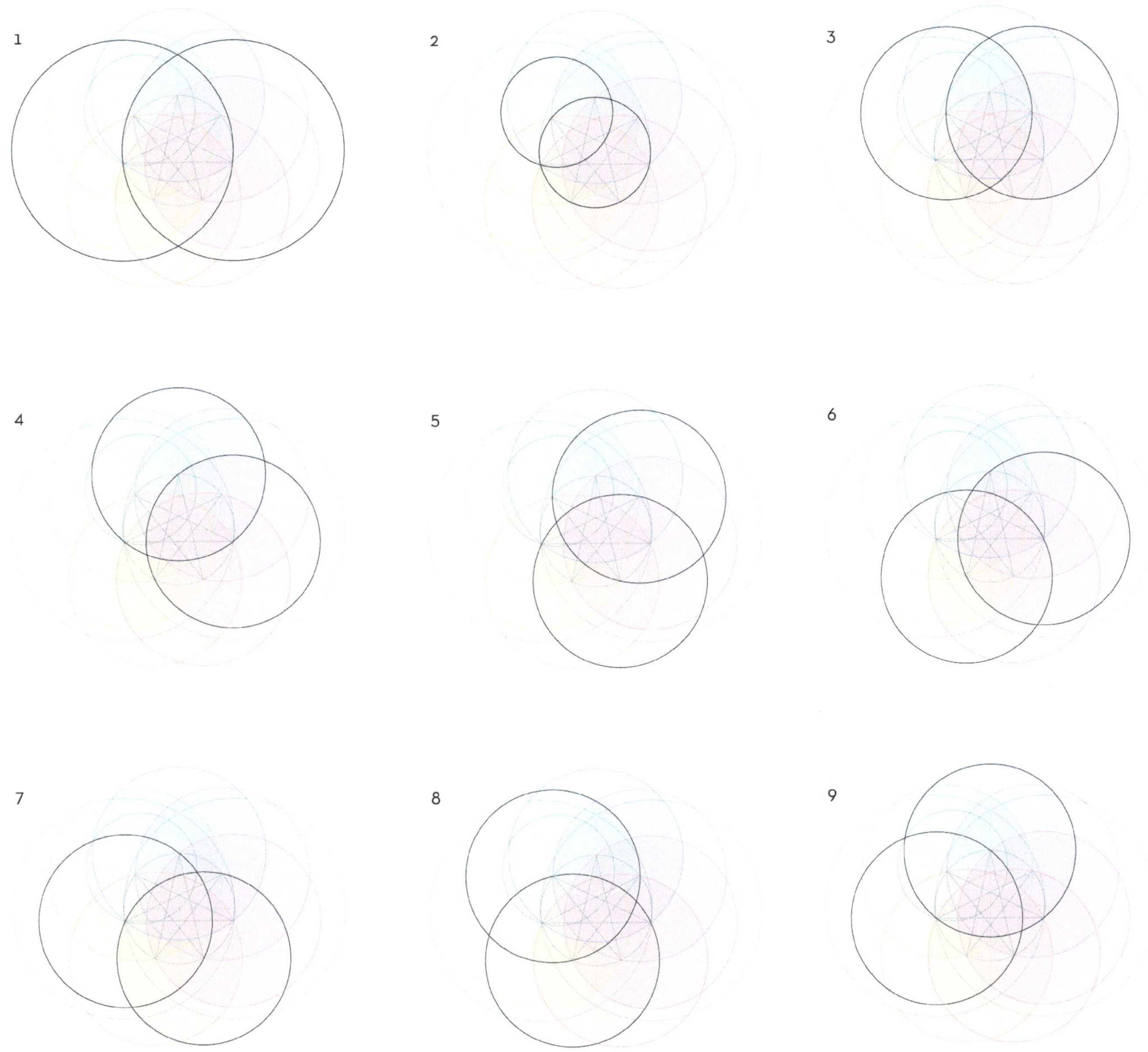

The construction is made up almost entirely of the Vesica Piscis form, repeated nine times.

THE MATHEMATICS OF THE CHESTAHEDRON

The full mathematics of the Chestahedron are known, courtesy of Ronald Milito.
Scans of Ronald's original work are included below.

FORMULAS FOR THE CHESTERHEDRON*

by Ronald Milito 2010, March 9

P. A1

Short diagonal of kite

$$PQ = 0.5 - 1.5 \cos\theta$$

Long diagonal of kite

$$RT = \frac{0.57735}{\cos \omega} = \frac{0.57735}{\cos\left[\tan^{-1}\left(\frac{2\sin\theta}{1+\cos\theta}\right)\right]}$$

Area of a kite

$$Area = \frac{1}{2}(PQ)(RT) = \frac{0.1443 - 0.433\cos\theta}{\cos\left[\tan^{-1}\left(\frac{2\sin\theta}{1+\cos\theta}\right)\right]}$$

Dihedral angle of kite's long diagonal with the base (horizontal plane)

$$\omega = \tan^{-1}\left(\frac{SI}{IR}\right) = \tan^{-1}\left[\frac{0.866\sin\theta}{0.433(1+\cos\theta)}\right] = \tan^{-1}\left[\frac{2\sin\theta}{1+\cos\theta}\right]$$

Altitude of chesterhedron

$$TN = \frac{1.1547\sin\theta}{1+\cos\theta}$$

* [θ is the dihedral angle the opening equilateral triangular petal makes with the base.]

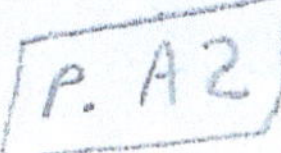

Distance from petal tip to top of chesterhedron's altitude

OR

Length of upper side of a chesterhedron kite (TH)

is given by

$$TH = \sqrt{(TL)^2 + (HL)^2}$$

where

$$TL = \frac{1.1547 \sin\theta}{1 + \cos\theta} - 0.8660 \sin\theta$$

and

$$HL = 0.28868 - 0.8660 \cos\theta$$

The angle TH makes with the horizontal plane is given by

$$\rho = \tan^{-1}\frac{TL}{HL} = \tan^{-1}\left[\frac{\frac{1.1547 \sin\theta}{1 + \cos\theta} - 0.8660 \sin\theta}{0.28868 - 0.8660 \cos\theta}\right]$$

The angle TH makes with the chesterhedron altitude is given by

$$\gamma = 180 - \rho$$

by Ronald Milito 2010, March 9

A.81

TABLE OF COMPARISON for chesterhedrons formed as three sides of a unit side tetrahedron open up as petals of a flower until they lie flat in a plane.

θ (theta) dihedral angle of petal with base	PQ length of kite's short diagonal	RT length of kite's long diagonal	Area of kite $\frac{1}{2}$(PQ)(RT)	ω (omega) dihedral angle of kite with base (RT with base)	TN altitude of the chesterhedron
70.5288° tetrahedral case	0	1	0	54.7356°	0.8165
90° tip of petal at maximum altitude of 0.8660	0.5	1.2910	0.3228	63.4349°	1.1547
94.8304° when a kite area equals a petal area	0.6263	1.3827	0.4330	65.3200°	1.2564
109.4712° when the petals have opened to octahedral position, the kites have become rhombuses, and ✩ the acute angle of the petals to the horizontal has become 70.5288°	1.0	1.7320	0.8660 the kite become rhombus is made of two equilateral triangles or petals each of area 0.433	70.5288° the same as the dihedral angle of the tetrahedron + ✩ it's also the closest angle of the petals to the horizontal.	1.6330
138.1898° when the petals have opened to icosahedral position	1.618 (the golden number)	3.0778	2.4899	79.1877°	3.0231
179°	1.9998	132.32	132.299	89.7500°	132.32
179.9°	1.99999	1,323.19	1,323.17	89.9750°	1,323.2
↓ 180° flattened out petals, and chesterhedron is stretching to infinity, + the plane figure formed is ~ equilateral triangle twice the size of the starting base.	↓ 2	↓ ∞	↓ ∞	↓ 90°	↓ ∞

θ (theta) dihedral angle of a petal with the base	TL	LH	TH length of an upper side of a chesterhedron kite	ρ (rho) angle of TH with the horizontal	γ (gamma) angle of TH with the altitude of the chesterhedron
70.5288° tetrahedral case no chesterhedron	0	0	0	—	—
90° tip of petal at maximum altitude of 0.8660	0.2887	0.2887	0.4083	45°	135°
94.8304° a kite area = a petal area	0.3935	0.3616	0.5344	47.4169°	132.5831°
109.4712° octahedral position kites are rhombuses	0.8165	0.5773	1.000	54.7381°	125.2619°
138.1898° icosahedral position	2.4457	0.9342	2.6180 (golden)	69.0951°	110.9049°
179°	132.300	1.1545	132.305	89.5000°	90.5
179.9°	1,323.188	1.1547	1,323.188	89.9500°	90.05
↓ 180°	↓ ∞	↓ 1.1547 $(\frac{2}{3}\sqrt{3})$	↓ ∞	↓ 90°	↓ 0°

Visualizing The Geometric Model For Developing Formulas For Various Measures Of The Chesterhedron.

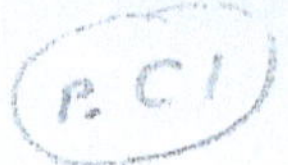

by Ron Milito 2010, March 9

Figure 1

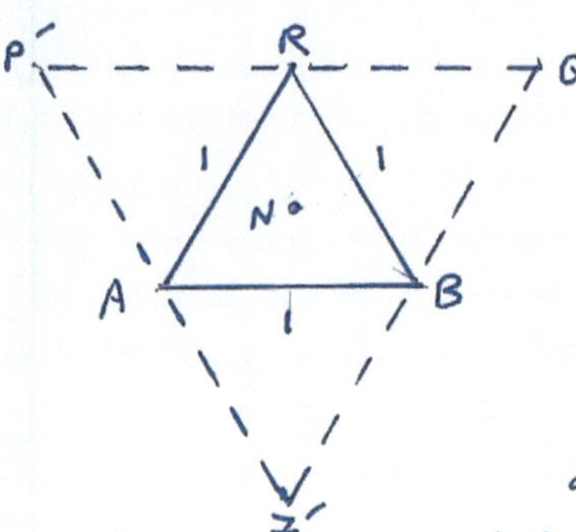

Picture a unit equilateral triangle, $\triangle ARB$, with center N.

$AR = RB = BA = 1$

Next surround $\triangle ARB$ with three equilateral triangles all congruent to it.

$\triangle AP'R \cong \triangle RQ'B \cong \triangle AZ'B \cong \triangle ARB$

Figure 2

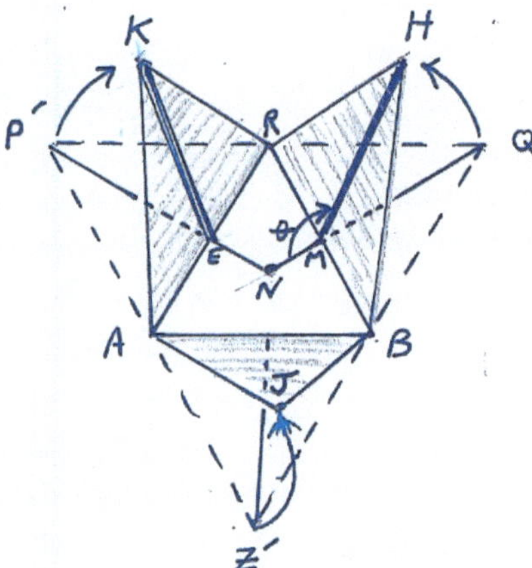

Picture triangles $AP'R$, $RQ'B$, $AZ'B$ as hinged petals rotating up from the plane becoming respectively triangles AKR, RHB, AJB

Please note that to start with we are viewing angle HMQ' as increasing, but we can also say that angle NMH is decreasing.

In our development of formulas we will focus on angle NMH and define it as the dihedral angle of a petal with its base.

θ = dihedral angle for a petal = $\measuredangle NMH$

Figure 3

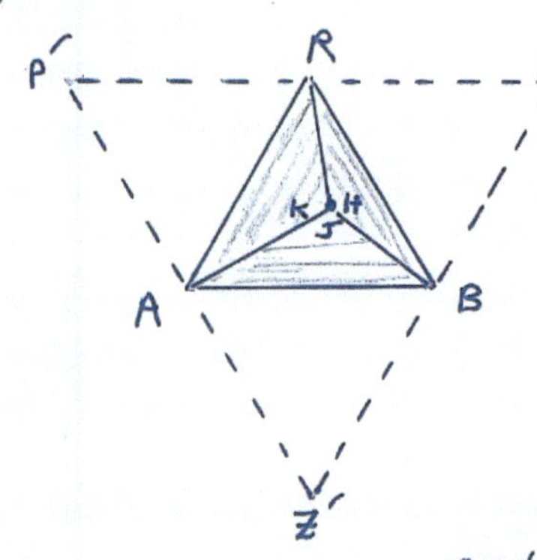

In Figure 3 we have an above view of the tetrahedron formed when the three petals meet.

Then $\theta = 70.5288° =$ (the dihedral angle for two sides of a tetrahedron)

and $\measuredangle HMQ' = 109.4712°$ which is the supplement of θ

Figure 4

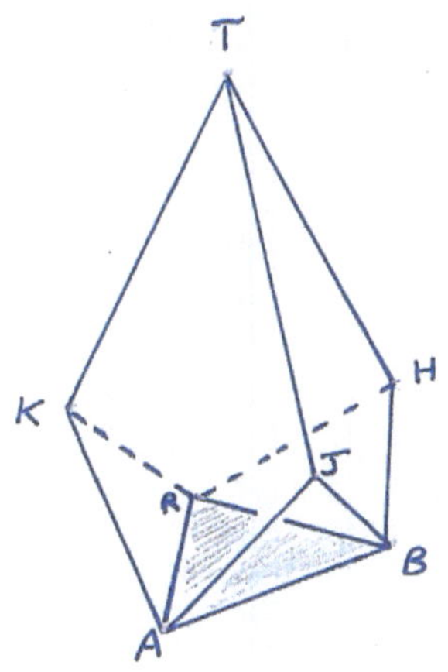

A chesterhedron has seven faces:
four congruent equilateral trianges
ΔARB (the base) & the petals
ΔAKR, ΔRHB, ΔAJB
and three congruent kites
AKTJ, BJTH, RHTK

As the petals open as depicted in figure 2, a series of chesterhedrons are generated.

Figure 5

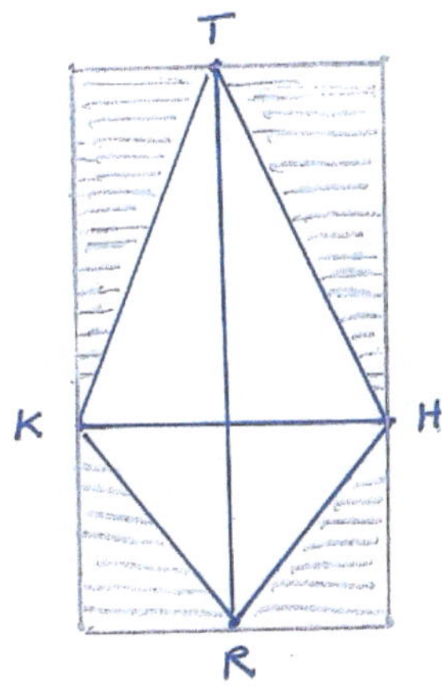

kite RHTK with
short diagonal KH
& long diagonal RT
encompassed by a rectangle
to illustrate that the
area of a kite = $\frac{1}{2}$(short diagonal)(long diagonal)

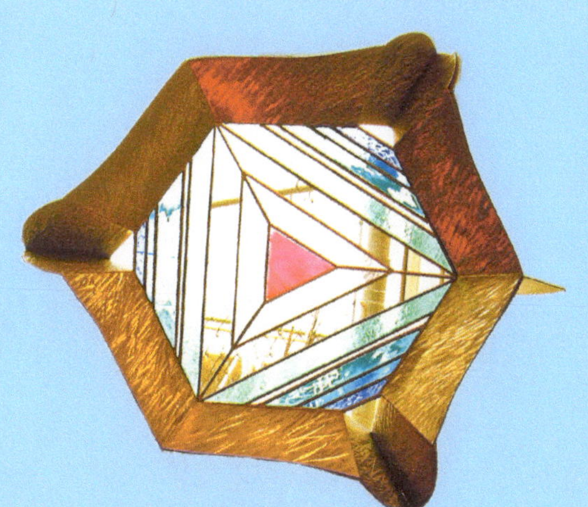

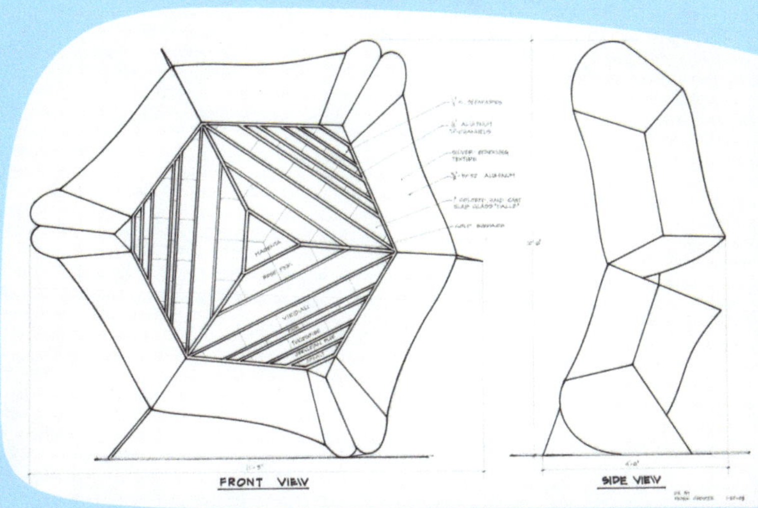

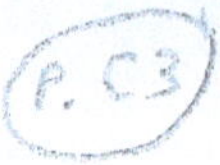

Figure 6 : A modification of Figure 2 that allows the derivation of the length of a kite's short diagonal KH as a function of θ

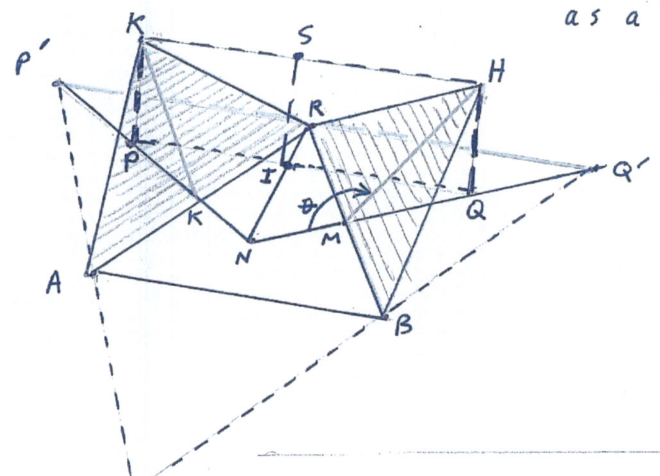

We see that PQ is the perpendicular projection of KH onto the plane of the base, ΔARB with S and I their corresponding midpoints

1.) $IQ = NQ \cos \measuredangle NQI$	1.) by trig
2.) $\measuredangle NQI = 30°$	2.) properties equilat triangles & right triangles
3.) $IQ = NQ \cos 30° = NQ\,(0.8660)$	3.) sub 2 into 1 & evaluate
4.) $PQ = 2\,IQ = 2\,(0.8660)\,NQ = 1.732\,NQ$	4.) I is a mid point of PQ; sub 3 into $PQ = 2\,IQ$ on line of & evaluate
5.) $NQ = NM + MQ$	5.) whole = sum of parts
6.) $NM = \frac{\sqrt{3}}{6} = 0.28868$	6.) properties of unit equilat triangle
7.) $MQ = MH \cos \theta$	7.) by trig
8.) $MH = 0.8660$	8.) properties unit equilat triangle
9.) $MQ = 0.8660 \cos \theta$	9.) sub 8 into 7
10.) $NQ = 0.28868 - 0.8660 \cos \theta$	10.) sub 6 & 9 into 5. A negative sign is needed because $\cos\theta$ is positive when $0° \leq \theta < 90°$ & $\cos\theta$ is negative when $90° < \theta \leq 180°$
11.) $IQ = 0.8660\,(0.28868 - 0.8660 \cos\theta)$	11.) sub 10 into 3
12.) $PQ = 2\,IQ = 2\,(0.8660)(0.28868 - 0.8660 \cos\theta)$	12.) sub 11 into 4
☆ 13.) $PQ = 0.5 - 1.5 \cos\theta$ kite short diagonal	13.) simplify & evaluate 12

P. C4

Figure 7: Figures 4 and 6 both rotated 120° with selected features combined in a chesterhedron

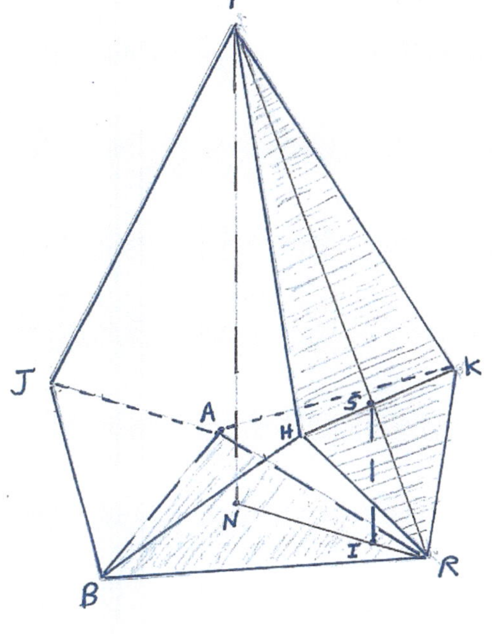

Figure 8: Triangles RNT and RIS abstracted from Figure 7

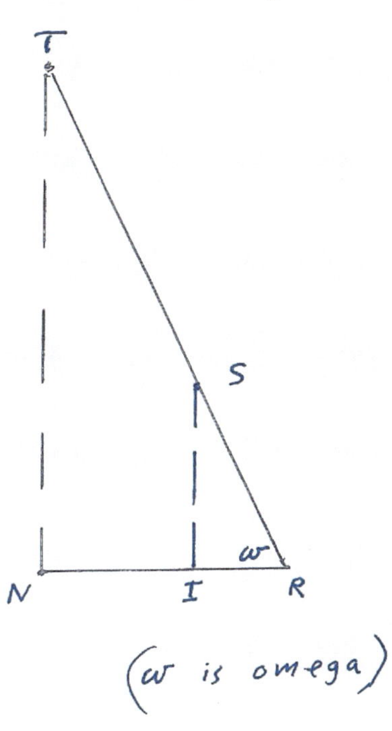

(ω is omega)

1.) $IR = NR - NI$ whole + parts

2.) $NR = 0.57735$ properties equilat triangle

3.) $NI = NQ \sin 30 = 0.5\ NQ$ by trig from Figure 6 + evaluation

4.) $NI = 0.5\,(0.28868 - 0.8660 \cos\theta)$ substitution for NQ from line 10 under Figure 6

5.) $NI = 0.1443 - 0.433 \cos\theta$ simplification + evaluation

6.) $IR = 0.57735 - (0.1443 - 0.433 \cos\theta)$ sub lines 2 + 5 into line 1

7.) $IR = 0.433\,(1 + \cos\theta)$ simplification + evaluation

---

8.) $SI = HQ = 0.8860 \sin\theta$ by trig from Figure 6

9.) $\omega = \tan^{-1}\left(\frac{SI}{IR}\right) = \tan^{-1}\left[\frac{0.8660 \sin\theta}{0.433\,(1+\cos\theta)}\right] = \tan^{-1}\left[\frac{2 \sin\theta}{1+\cos\theta}\right]$ ☆

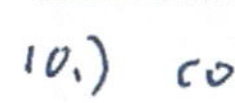

10.) $\cos \omega = \frac{NR}{RT}$ by definition

11.) $RT = \frac{NR}{\cos \omega} = \frac{0.57735}{\cos\left[\tan^{-1}\left[\frac{2\sin\theta}{1+\cos\theta}\right]\right]}$ ☆ kite long diagonal

---

12.) $\frac{TN}{NR} = \frac{SI}{IR}$ by similar triangles

13.) $TN = NR\left(\frac{SI}{IR}\right) = 0.57735\left(\frac{0.8660 \sin\theta}{0.433(1+\cos\theta)}\right) = \frac{1.1547 \sin\theta}{1 + \cos\theta}$ ☆ altitude of chesterhedron

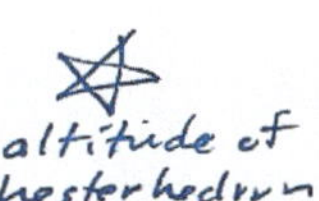

Figure 9: Combination of aspects from Figures 7 and 6 allowing calculation of the upper sides of a kite (see Figure 4 — TK, TJ, TH)

and the angle ρ such an edge makes with the horizontal

or the angle γ with the altitude TN

all as functions of θ

(ρ is rho)

(γ is gamma)

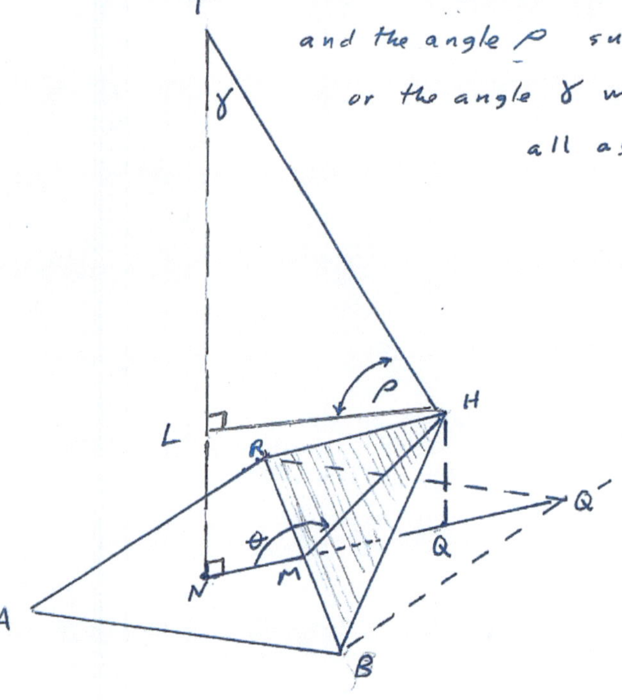

1.) $TL = TN - LN = TN - HQ = TN - SI = \dfrac{1.1547 \sin\theta}{1 + \cos\theta} - 0.8660 \sin\theta$

by trig see Figure 6 & Figure 7 lines 8 & 13

2.) $HL = NQ = 0.28868 - 0.8660 \cos\theta$ see Figure 6 line 10

3.) $TH = \sqrt{(TL)^2 + (HL)^2}$ by Pythagoras

4.) $\tan\rho = \dfrac{TL}{LH}$ by definition

5.) $\rho = \tan^{-1}\left(\dfrac{TL}{LH}\right) = \tan^{-1}\left[\dfrac{\dfrac{1.1547 \sin\theta}{1 + \cos\theta} - 0.8660 \sin\theta}{0.28868 - 0.8660 \cos\theta}\right]$

6.) $\gamma = 90^\circ - \rho$

Finding The Petal Dihedral Angle For A Chesterhedron With A Kite Area Equal To A Petal Area Making All Seven Faces Have Equal Area

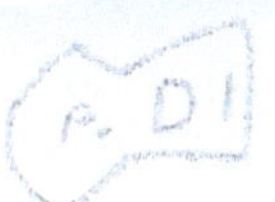

by Ronald Milito 2010, March 9

I programmed a calculator using a formula I derived expressing a kite's area as a function of a petal's dihedral angle with the base. Since the area of an equilateral triangle with unit side (equal to one) is 0.433, I made trial entries closing in this value. Below are my trials.

θ	Kite Area	
90°	0.323	too low
95°	0.437	too high
94°	0.413	too low
94.5°	0.425	too low
94.6°	0.427	too low
94.8°	0.4323	too low
94.9°	0.4347	too high
94.85°	0.43348	too high
94.83°	0.43299	almost but low
94.84°	0.4332	too high
94.831°	0.43302	slightly high
94.8305°	0.43301	closer but still high
94.8304°	0.43300	the value I accepted.

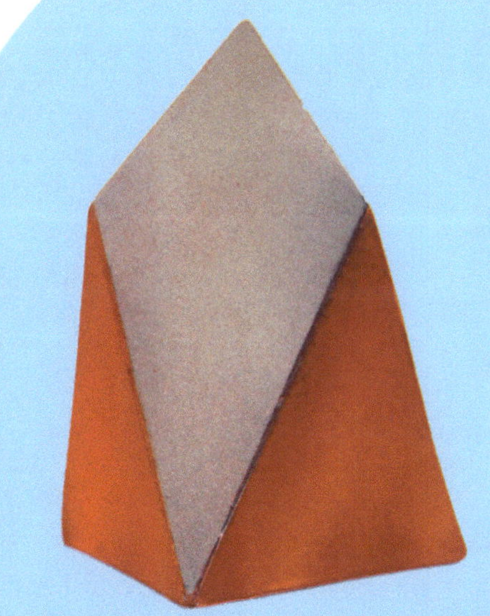

MATHEMATICAL SUMMARY

The dihedral angle Theta is the angle between the base triangle ABC with any of the other three upper triangles (the "petals" of the opening Tetrahedron).

Below are the X-Y-Z Chestahedron coordinates with 0,0,0 as the base triangle's center, and with a side of length = 1.
(X Axis = left-right, Y axis = up-down, Z axis = front-back)

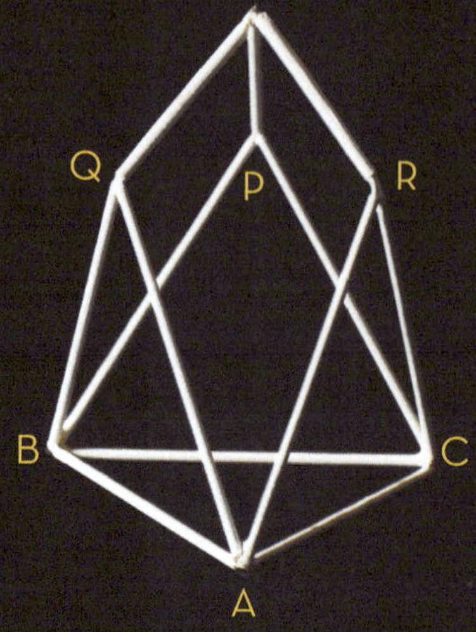

Base Triangle, Vertices ABC		
Ax, with Ay = 0, Az = 0	Bx, with By = 0, Bz = - 0.50	Cx, with Cy = 0, Cz = 0.50
0.577350269	-0.288675135	-0.288675135

Upper Triangles, Vertices PQR		
Px	Py	Pz
-0.361608072	0.86294889	0
Qx	Qy	Qz
0.180804036	0.86294889	-0.313161776
Rx	Ry	Rz
0.180804036	0.86294889	0.313161776

Dihedral angle Theta:	94.83092618 °
Fraction of Theta in Minutes:	49.8555709 '
Fraction of Theta in Seconds:	51.33425387 "
Theta in radians:	1.655111895 rad
Dihedral angle Omega (base triangle to kite):	65.32005574 °
Angle Omega in radians:	1.14005004 rad

Additional Calculations:	
Height (Apex point T) Y-coordinate only, Tx=Tz= 0	1.256407783
PQ = QR =RQ = Horiz of Kite	0.626323553
AT=BT=CT= Vertical of Kite	1.382712498
Area of any Equilateral Triangle (fixed)	0.433012702
Area of Kite = 0.5 x AT x PQ	0.433012702
Total Surface Area of Chestahedron (4 x Triangle + 3 x Kite SA)	3.031088913
Length of top edges of Kite = PT=QT=RT (Bottom edges of the Kite always = 1)	0.534387779

Angles inside of Kite:	
Angle Alpha (PTQ) at top of Kite	71.75012248 °
Angle Beta (PBQ) at bottom of Kite	36.49975504 °
Side Angles of Kite at Vertices P, Q, or R	125.8750612 °

Radius of Circle (r1) enclosing Base Triangle	0.577350269
Radius of Circle enclosing Triangle PQR	0.361608072
Radius of Circle circumscribed by Base Triangle	0.288675135

Angle Theta for octahedral opening	109.4712204
Angle Theta for Height being Phi exactly	108.9733377

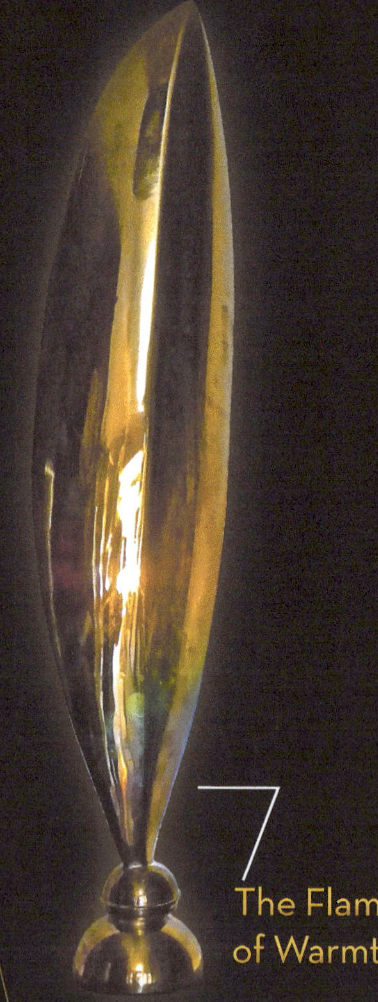

The Flame of Warmth

The mathematics has been independently confirmed by Dr. Karl Maret.

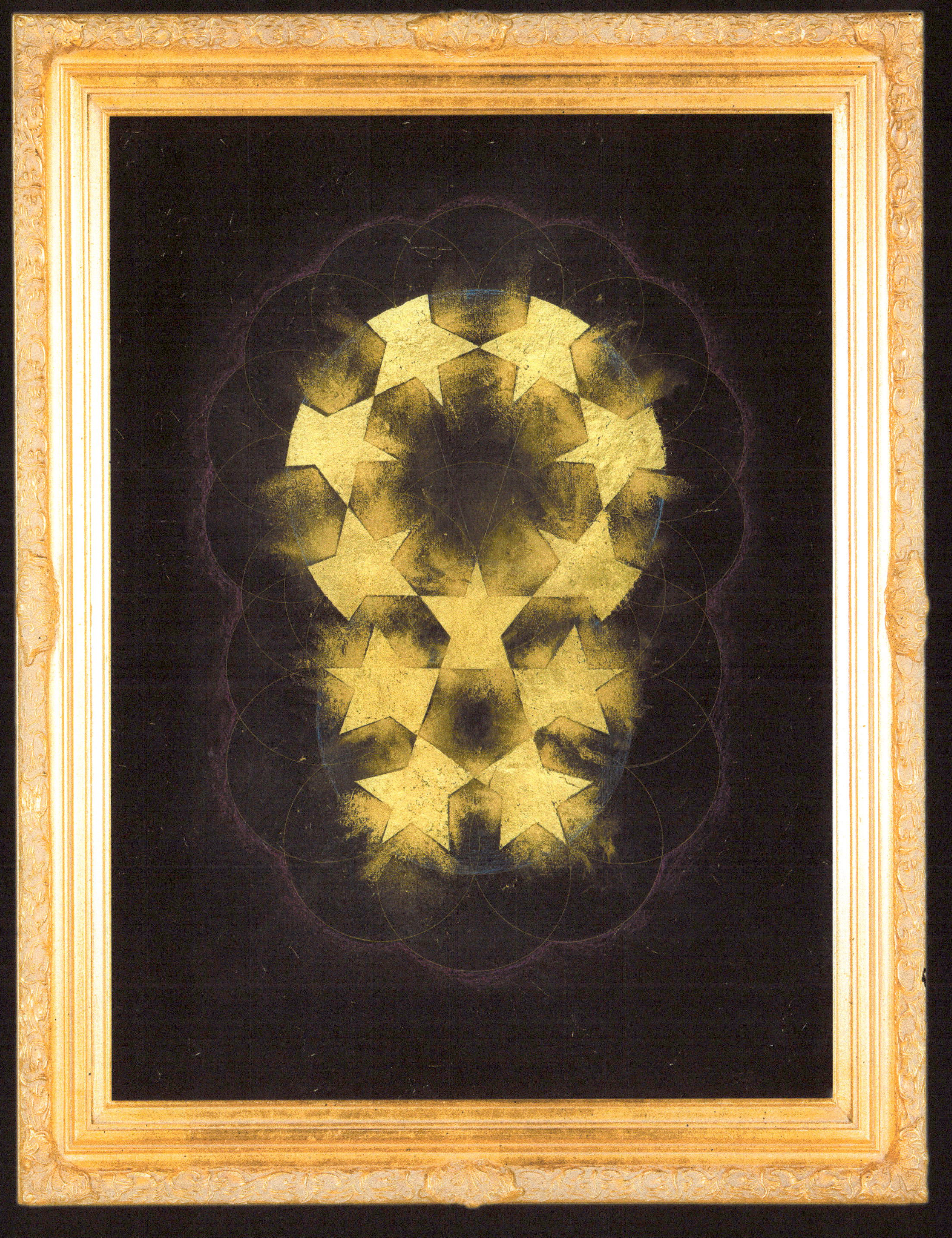

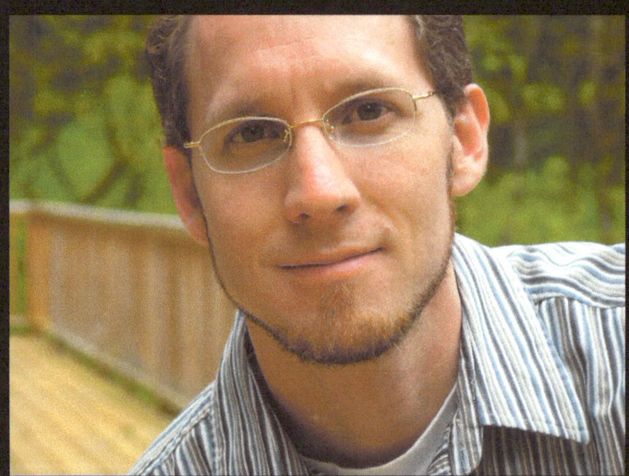

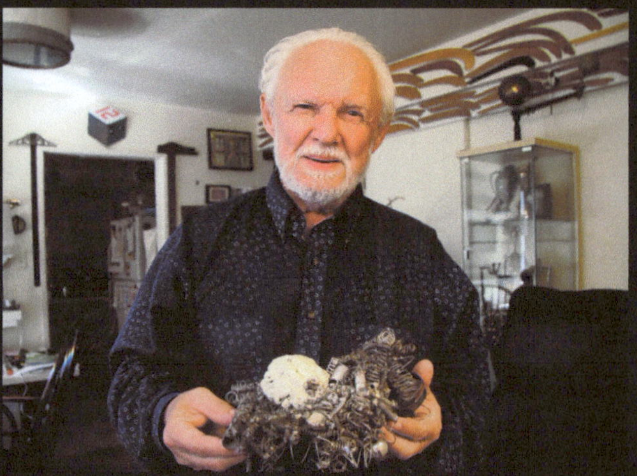

About the Author

Seth Miller has a closet of hats: professor, Waldorf teacher, designer, author, tech support, webmaster, and perpetual student. He has a BA in philosophy from The Colorado College, an MA in consciousness studies from JFK University, and is currently writing his PhD dissertation exploring epistemological connections between anthroposophy and second-order cybernetics. It's as esoteric as it sounds, but he is sure you'd find it fascinating over a nice cup of coffee sometime.

About the Artist

Frank Chester is an artist, sculptor, teacher, and geometrician based in San Francisco. Since encountering the work of Rudolf Steiner, Frank has been exploring the relation between form and spirit.

Frank has been recognized around the world for his unusual melding of art and science, and for his dynamic and inspiring presentations.

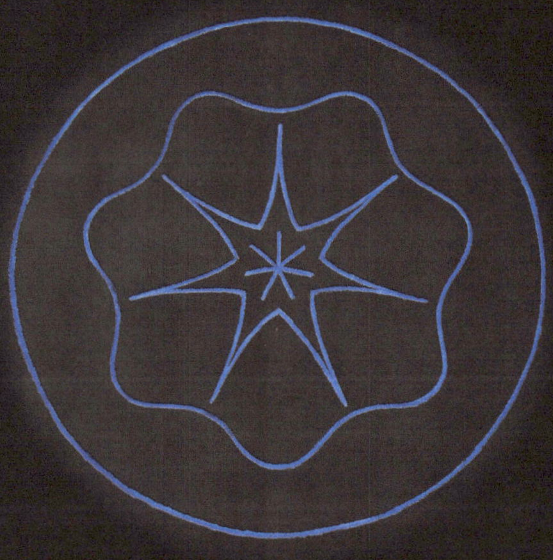

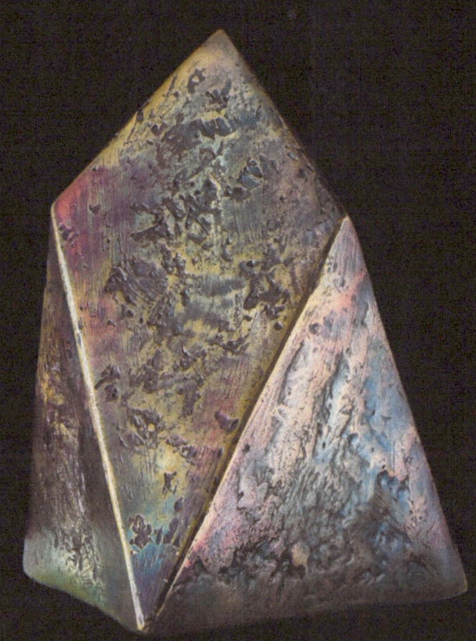

SpiritAlchemy.com

Video presentations and much more at

FrankChester.com

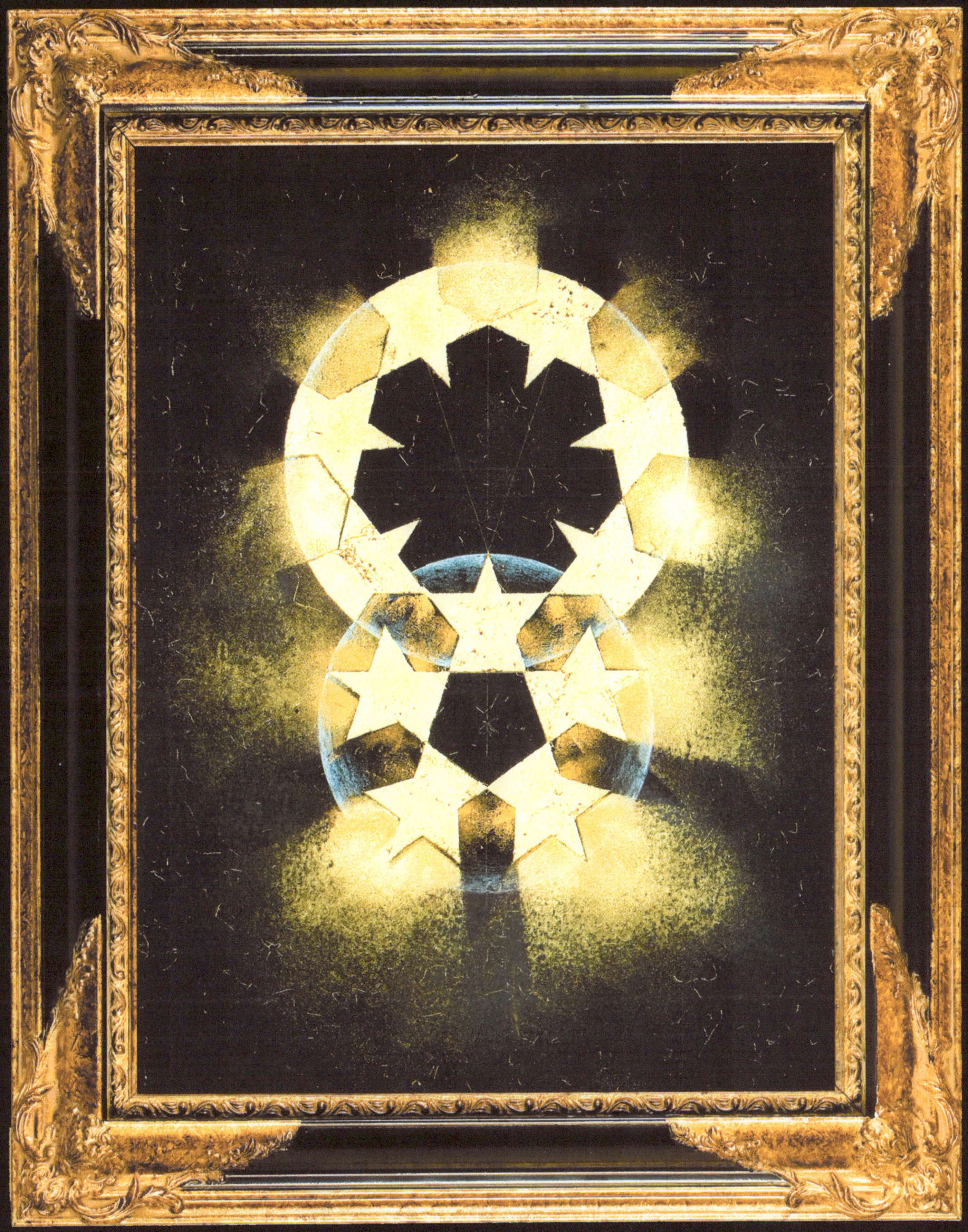

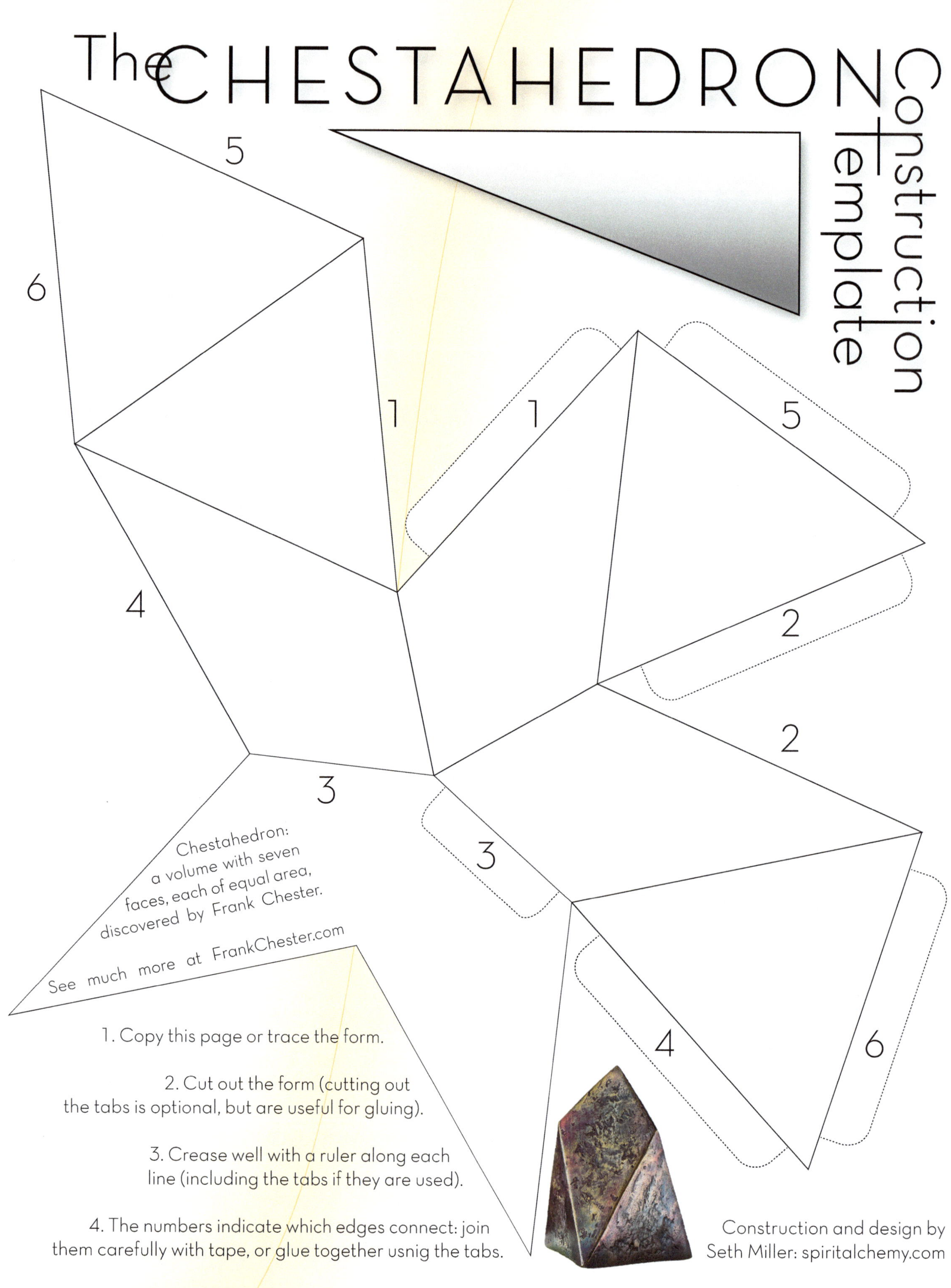
The CHESTAHEDRON
Construction template
5
6
1
1
5
4
2
2
3
3
Chestahedron: a volume with seven faces, each of equal area, discovered by Frank Chester.
See much more at FrankChester.com
1. Copy this page or trace the form.
2. Cut out the form (cutting out the tabs is optional, but are useful for gluing).
3. Crease well with a ruler along each line (including the tabs if they are used).
4. The numbers indicate which edges connect: join them carefully with tape, or glue together usnig the tabs.
4
6
Construction and design by Seth Miller: spiritalchemy.com

www.ingramcontent.com/pod-product-compliance
Lightning Source LLC
LaVergne TN
LVHW071633100826
845154LV00008BA/141

* 9 7 8 0 9 8 8 7 4 9 2 0 7 *